T0207251

Communications in Computer and Information Science 1767

Rationale
The CCIS series is devoted to the publication of proceedings of computer science conferences. Its aim is to efficiently disseminate original research results in informatics in printed and electronic form. While the focus is on publication of peer-reviewed full papers presenting mature work, inclusion of reviewed short papers reporting on work in progress is welcome, too. Besides globally relevant meetings with internationally representative program committees guaranteeing a strict peer-reviewing and paper selection process, conferences run by societies or of high regional or national relevance are also considered for publication.

Topics
The topical scope of CCIS spans the entire spectrum of informatics ranging from foundational topics in the theory of computing to information and communications science and technology and a broad variety of interdisciplinary application fields.

Information for Volume Editors and Authors
Publication in CCIS is free of charge. No royalties are paid, however, we offer registered conference participants temporary free access to the online version of the conference proceedings on SpringerLink (http://link.springer.com) by means of an http referrer from the conference website and/or a number of complimentary printed copies, as specified in the official acceptance email of the event.

CCIS proceedings can be published in time for distribution at conferences or as post-proceedings, and delivered in the form of printed books and/or electronically as USBs and/or e-content licenses for accessing proceedings at SpringerLink. Furthermore, CCIS proceedings are included in the CCIS electronic book series hosted in the SpringerLink digital library at http://link.springer.com/bookseries/7899. Conferences publishing in CCIS are allowed to use Online Conference Service (OCS) for managing the whole proceedings lifecycle (from submission and reviewing to preparing for publication) free of charge.

Publication process
The language of publication is exclusively English. Authors publishing in CCIS have to sign the Springer CCIS copyright transfer form, however, they are free to use their material published in CCIS for substantially changed, more elaborate subsequent publications elsewhere. For the preparation of the camera-ready papers/files, authors have to strictly adhere to the Springer CCIS Authors' Instructions and are strongly encouraged to use the CCIS LaTeX style files or templates.

Abstracting/Indexing
CCIS is abstracted/indexed in DBLP, Google Scholar, EI-Compendex, Mathematical Reviews, SCImago, Scopus. CCIS volumes are also submitted for the inclusion in ISI Proceedings.

How to start
To start the evaluation of your proposal for inclusion in the CCIS series, please send an e-mail to ccis@springer.com.

Victor Taratukhin · Artem Levchenko ·
Yury Kupriyanov
Editors

Information Systems and Design

Third International Conference, ICID 2022
Tashkent, Uzbekistan, September 12–13, 2022
Revised Selected Papers

Editors
Victor Taratukhin
University of Muenster
Münster, Germany

Artem Levchenko
SAP SE
Tokyo, Japan

Yury Kupriyanov
SAP SE
Walldorf, Germany

ISSN 1865-0929 ISSN 1865-0937 (electronic)
Communications in Computer and Information Science
ISBN 978-3-031-32091-0 ISBN 978-3-031-32092-7 (eBook)
https://doi.org/10.1007/978-3-031-32092-7

This Springer imprint is published by the registered company Springer Nature Switzerland AG
The registered company address is: Gewerbestrasse 11, 6330 Cham, Switzerland

Paper in this product is recyclable.

Preface

We are delighted and honored to present to you the proceedings of the Third International Conference for Information systems and Design (ICID 2022) which took place September 12–13, 2022, in Tashkent, Uzbekistan.

As stated in the original mission of the conference: ICID is set to become a global cooperation network that promotes open industry innovations in the academic environment. We focus on practical results-based studies prepared by academic researchers and industry experts in Information Systems design, deployment, and adoption. The ICID community also welcomes students at its conference to present their research and participate in the experimental workshops (e.g., Ideathons, Hackathons, and the like).

The main value of ICID community development is seen in building global connections between academia and business. The ICID conference is an active international scientific event with 100+ participants and guests from Germany, Central Asia, Japan, South Korea, UK, Finland, Belgium, the USA, and other countries.

In 2022 the organizational committee reviewed 35 and accepted 12 of the submitted papers. Based on the clear criteria of research focus, depth, and results application a decision was made to split accepted submissions into two sections, listed below, and thus ensure the proper positioning of the authors within the ICID community.

The two sections of ICID 2022 papers are

- Methodological Support of analysis and management tools: theoretical-focused research
- Digital transformation of enterprises based on analysis and management tools: practical-focused research

The Third ICID conference was run preliminary in the off-line/on-site format and was attended by 70 people in-person and 40 participants joining on-line via videoconferencing. Following the ICID 2022 program, the conference took place over 2 days with plenary sessions on day 1, with 26 presentations on the first day and 14 presentations on the 2nd.

The conference on-site participants included leading experts and teams from key Central Asian universities, research centers, and leading companies in IT and industry.

September 2022

Victor Taratukhin
Artem Levchenko
Yury Kupriyanov

Preface

Organization

International Chair

Jörg Becker	University of Muenster, Germany

General Chairs

Victor Taratukhin	University of Muenster, Germany
Mikhail Matveev	Voronezh State University, Russia

Industry Chairs

Yury Kupriyanov	SAP SE, Germany
Artem Levchenko	SAP SE, Japan

International Program Committee

Jörg Becker	University of Muenster, Germany
Victor Taratukhin	University of Muenster, Germany
Yury Kupriyanov	SAP SE, Germany
Artem Levchenko	SAP SE, Japan
Muzaffar Djalalov	INHA University, Uzbekistan
Mikhail Matveev	Voronezh State University, Russia
Soh Kim	Stanford Center at the Incheon Global Campus, South Korea
Sergey Balandin	The FRUCT Association, Finland

Organizing Committee

Victor Taratukhin	University of Muenster, Germany
Yury Kupriyanov	SAP SE, Germany
Artem Levchenko	SAP SE, Japan
Dilshoda Mirzokhidova	INHA University, Uzbekistan

Yulia Bolshakova — National Research University Higher School of Economics, Russia
Yulia Skrupskaya — University of Lapland, Finland

Contents

Methodological Support of Analysis and Management Tools: Theoretical-Focused Research

Development of the Simulation Model of Complex Evaluation of the Quality of the Emergency Medical Services

Svetlana Begicheva[1](✉) and Antonina Begicheva[2]

[1] Ural State University of Economics, 8 Marta 62/45, 620144 Yekaterinburg, Russia
begichevas@mail.ru

[2] National Research University Higher School of Economics, Pokrovsky Boulevard 11, 109028 Moscow, Russia

Abstract. The research is dedicated to the questions of development of the simulation model of the complex evaluation of the quality of the emergency medical services. The author suggests supplementing the information as a medical assistance, which according to the method of A. Donabedian can be taken from three categories: "the quality of the structure', "the quality of the process" and "the quality of the results", the information from the category "the quality of the living environment". On the basis of the suggested conceptual model of the quality of the emergency medical services is defined the model of modelling activity EMS, which includes the submodels, the total amount of which will allow to take into account all the categories of the quality of the emergency medical services. There is an example of the model realization, which conjoins the combination of the methods of system dynamics and agent modelling. The multi-approach simulation model of the ambulance activity, developed with this concept in mind, takes into account the specifics of the EMS in a metropolis and allows one to assess the indicators of various categories of EMS quality.

Keywords: The model of Donabedian · simulation modelling · multimethod modelling · emergency medical services · the quality of the medical services

1 Introduction

One of the priority tasks of the State program of Russian Federation "The Development of the Public Health Services" [6] is called the increase of accessibility and the quality of the emergency medical services. Insufficient information on the field of the matters connected with the problems of building of models of complex evaluation of the quality of emergency medical services provides the urgency of the research.

Regulatory and legal framework of Russian Federation in the field of public health services contains the following definition: "the quality of the medical services is the summation of the characteristics which reflect the timeliness of the provision of medical care, the correct choice of methods of prevention, diagnosis, treatment and rehabilitation providing the medical care, the degree of achievement of the planned result [8].

V. Taratukhin et al. (Eds.): ICID 2022, CCIS 1767, pp. 3–14, 2023.
https://doi.org/10.1007/978-3-031-32092-7_1

The main criteria for the quality of medical care used in medical institutions of the Russian Federation include: the availability of medical care; adequacy, expressed in accordance with the technology of medical services to the needs of the population; efficiency and effectiveness; patient orientation and satisfaction; safety of the treatment process; timeliness of medical care; absence (minimization) of medical errors; scientific and technical level of organization of the provision of emergency medical care [17]. To assess medical care, the so-called quality indicators are used – numerical indicators that indirectly reflect the quality of its main components: structure, processes and results [9]. The principles of quality in the healthcare sector are also applicable to the assessment of the quality of emergency medical care.

When planning the activities of an ambulance, it is important not only to determine the quality of its work in the past, but also, based on data on the current state, to calculate the expected quality of service in order to identify the need for additional modernization measures. A typical way to determine the expected quality of service is through simulation.

Simulation modeling allows you to display the dynamics of the behavior of a system consisting of a large number of interacting interdependent components. The need to use simulation modeling in solving problems of quality management and the availability of EMS is justified by L.F. Laker and others. In the article "Understanding emergency care delivery through computer simulation modeling» [11] the authors provide a classification of existing studies and indicate that most of the works are devoted to modeling such processes of EMS activity as regulating patient flows, assessing the cost of services, managing ambulance teams, resource planning, etc. The target variables are most often two or more quality indicators from the following list: the average time for the team to reach the place of call, the average time of a patient's stay in the emergency department, the number of calls served, and resource allocation indicators [3]. L.F. Laker and others [11] formulated directions for future research in the field of simulation modeling of the EMS and recommended using the parameters of the patient's health status as target indicators.

The article will present an approach to the simulation of the EMS activity, based on the provisions of the theory of the doctor and the founder of the scientific study of quality in health care A. Donabedian. The proposed modification of the classical model will make it possible to assess the impact of managerial decisions on changes in the structure of the organization and the process of providing emergency medical services on the availability and quality of emergency medical care in the context of the growth of megacities and the development of their infrastructure.

2 Methods

The main goal of the modernization of the ambulance delivery system is to increase the availability and quality of ambulance services. The target indicators of the EMS work fit into the classical methodological scheme for assessing quality in health care, proposed by A. Donabedian [7] in 1988. According to his scheme, it is customary to distinguish three main components of the quality of medical care:

1. the quality of the structure of the organization of treatment,

2. the quality of the process of providing medical care,
3. the quality of the results.

A. Donabedian defined the quality of the structure as a component of the quality of medical care, describing the conditions for the provision of medical care: material, professional, organizational and managerial resources of a medical organization. The process includes all manipulations to help the patient from the moment of diagnosis to the end of treatment. The result of the provision of medical care is understood as the ratio of the achieved results of treatment to the planned ones. A. Donabedian suggested that there is a unidirectional relationship between the indicators of the quality of medical care: the quality of the structure ensures the quality of the process of providing care, which in turn contributes to the quality of the treatment result. The conceptual model of quality management of medical care according to Donabedian is often presented in the form of a diagram of three blocks connected by arrows:

"structure quality" → "process quality" → "result quality"

It should be noted that despite the fact that Donabedian's approach is the dominant paradigm in assessing the quality of medical services in the health care sector, there are studies that indicate the shortcomings of the model. Thus, a sequential linear transition from structure to process and further to the result is recognized as too simplified and does not allow to fully take into account the mutual influences and interactions between the components of the model [4, 12]. In addition, according to critics, the model does not take into account the influence of external factors [4, 5]. The influence of external factors on the structure of the organization is taken into account in the modification of the Donabedian model proposed by Nuckols and other authors [15]. The external factors influencing the quality and cost of health care, which were identified by the authors, are as follows: clinical indications for treatment, demographic characteristics and other determinants of health, including location, climate, employment opportunities, etc.

For a comprehensive analysis of the quality of emergency medical care, we propose to supplement the conceptual scheme proposed by A. Donabedian with such a component as the quality of the living environment. The indicators of the quality of the living environment are the dynamics of population indicators, transport infrastructure, demographic characteristics of the population, social infrastructure, social parameters of society, etc. [16]. The proposal is based on the works of Russian scientists [10, 14] and others, indicating that the organization of the functioning of the health care system is determined by such factors as population density, age-sex composition and health of residents, the socio-economic situation in the region, its natural and geographical peculiarities. A US Department of Transportation study [13] provides the results of a comparative analysis of ambulance activities in rural and urban areas, and notes that the configuration of EMS services is largely determined by the size, geographic, demographic and political location of the service region. Thus, the above factors influencing the formation of the structure of the organization of emergency medical care, affect the process of providing emergency medical services, and indirectly on the quality of the result.

Considering that the process of providing emergency medical care consists of two sequential subprocesses: (1) the process of receiving complaints from the population by the dispatcher and (2) the process of providing medical care by the ambulance team, we present a conceptual model for assessing the quality of ambulance as follows (see Fig. 1).

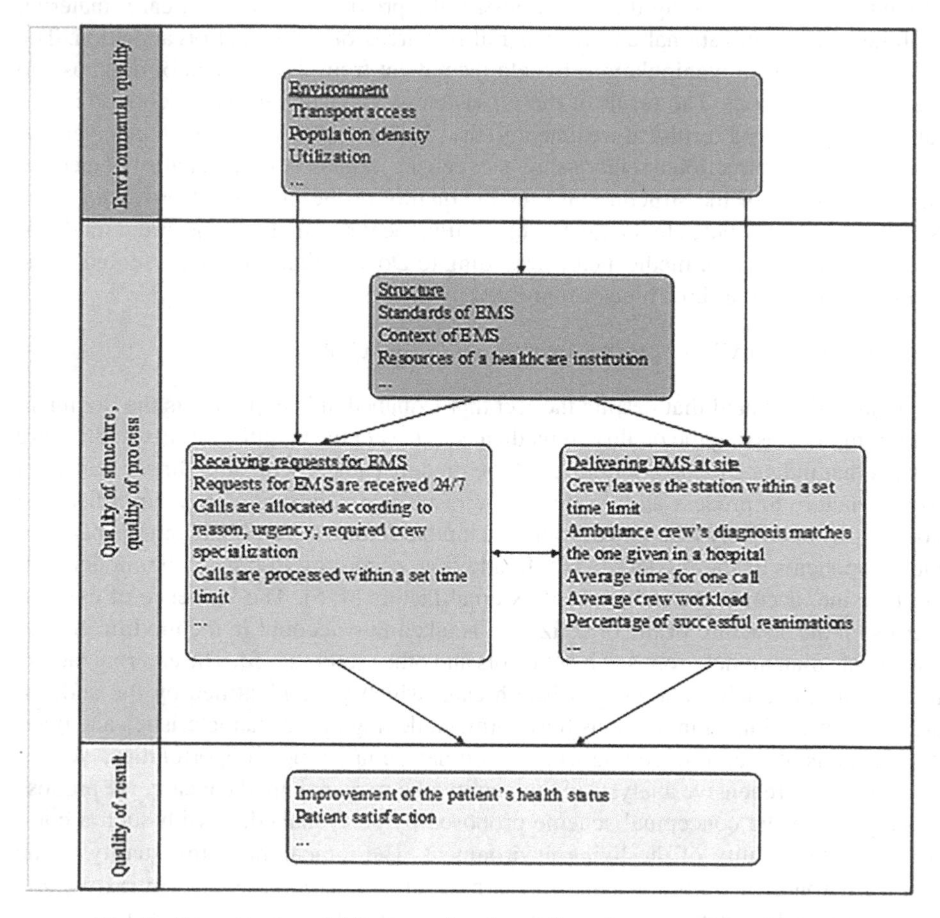

Fig. 1. Conceptual model for assessing the quality of emergency medical care

According to the proposed conceptual model, the quality of the result of the provision of emergency medical care is directly influenced by the type of settlement, which characterizes the living environment of those seeking emergency medical care. Despite the fact that the organization of ambulance for the population of all types of administrative-territorial units is based on the same principles, the existing differences between the types of settlements leave an imprint on the nature of the provision of emergency medical care. Such features of megalopolises as high growth rates of megalopolises; abrupt trends in population change; uneven building density; the rate of development of transport communications; the presence of many artificial boundaries that reduce the connectivity of the transport network; the low level of transport accessibility of the territories of the outskirts of the metropolis, form a specific list of tasks that determine the planning horizon of the ambulance activities.

In accordance with the proposed conceptual model, we will define the concept of modeling the EMS activity, which consists in identifying three objects – submodels, the totality of which will allow to comprehensively take into account all categories of ambulance quality (see Fig. 2).

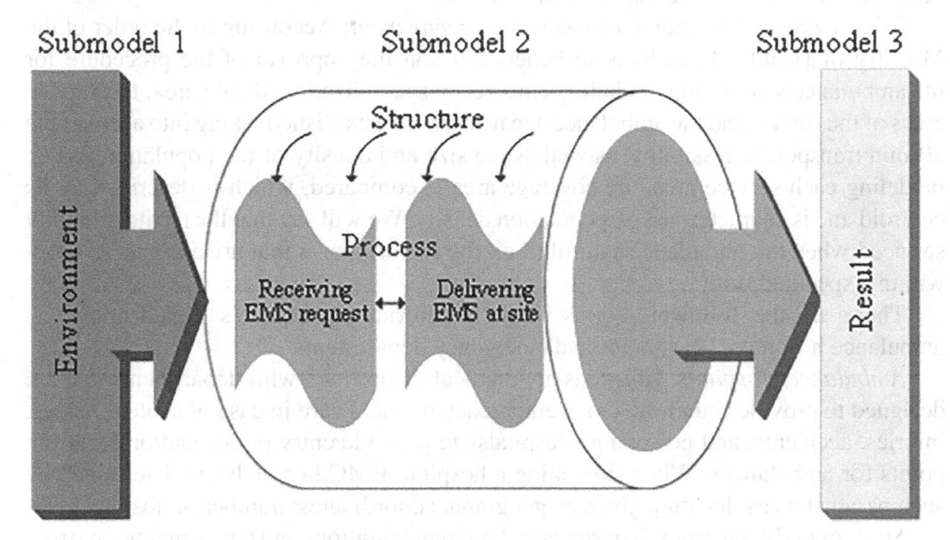

Fig. 2. Conceptual scheme of modeling activities for servicing ambulance calls

Submodel 1. The quality of the living environment affects the quality of life and health of the population, which, in turn, determine the need for ambulance services. The dynamics of changes in the indicators of the external flow can be taken into account when predicting the need for ambulance services based on the analysis of the influence of a set of reasons that form the number of calls to the emergency medical service. The system under study is described in terms of storage devices and flows between storage devices. The modeling method is system dynamics.

Submodel 2. Incoming calls to the EMS service, the number of which is predicted by model 1, are sequentially serviced by the dispatcher and the EMS team. Thus, submodel 2 is necessary to assess the quality of the organization's structure and service processes, which is the result of the interactions of the internal flows of the logistics system for the provision of EMS. When modeling the call service processes, one should take into account the importance of tracking interactions between model agents.

Submodel 3. During the operation of submodel 2, statistics are collected on performance indicators to assess the quality of the ambulance service. Indicators such as the time to reach a call, the average daily workload of the team, etc. are easily measured, but at the same time implicit characteristics of improving the health of patients. The third submodel is designed to assess the change in the quality indicators of the result when influencing the structure of the organization of treatment, while the quality of treatment results we propose to evaluate in such key indicators of the quality of medical care as patient satisfaction, receiving care in accordance with the reason for seeking treatment, and improving the health status of patients. The calculation of the outcome quality score (D) is based on aggregate patient scores. To calculate the quality indicators, the method of fuzzy assessment of the ambulance quality proposed in [1] is used.

Let us give a logical and mathematical description of the activity of servicing ambulance calls, including the development of a conceptual model and its formalized description. Let's describe the modeled objects.

Service area of the station or emergency department: According to the order of the Ministry of Health of the Russian Federation «on the approval of the procedure for the ambulance's provision, including emergency specialized medical care», the service areas of the station and the ambulance department are established taking into account the 20-min transport accessibility, as well as the size and density of the population. When modeling each service area, the coverage area is compared, which is described by its centroid and is characterized by population density. We will say that the territory is fully serviced when the ambulance can fulfill all the requirements that arise in this territory within a specified time.

There are the following forms of the territorial ambulance's organization: (1) ambulance hospitals; (2) stations and emergency departments.

Ambulance Hospitals: Hospitals are medical institutions with departments that are designed to provide round-the-clock emergency medical care in case of acute illnesses, injuries, accidents, and poisoning. Hospitals are possible entry points and/or departure points for ambulances. When simulating a hospital, EMC hospitals are determined by such parameters as: location given by geographic coordinates; number of hospital beds.

Stations and emergency departments: Ambulance stations and departments are structural elements of the ambulance service, the location of which is determined by 20-min transport accessibility to each point in the service area. Near the station, the EMC branch, there is an organized parking lot where ambulances can park while waiting for calls. Thus, the following characteristics of stations and branches of the EMC are important in the study: location given by geographic coordinates; number of mobile crews.

Calls are received and serviced by ambulance workers.

The paramedic for receiving ambulance calls and transferring them to mobile crews determines the urgency and profile of the call and decides to send a free mobile crew of the required profile to the patient (victim). Special software facilitates the work of the dispatch service and is used to control the number of crews on the line, the cars' location, determine the calls' priority, control the time of call execution.

The mobile ambulance crew immediately travels to the place where the ambulance is called, provides ambulance services based on the medical care's standards, takes measures to stabilize or improve the patient's condition and carry out a medical evacuation of the patient in the medical indications' presence. There are: general-purpose crews; specialized crews (psychiatric, pediatric, intensive care).

Patients are the main consumers of emergency medical services. In the context of research, patient calls can be classified according to:

- the area of the call;
- call priority (emergency, urgent);

- call profile (general, pediatrics, psychiatry, resuscitation);
- call time (day/night, a season of the year).

Let us describe the simulated processes.

At each station at the zero point in time, there is a number of crews: general and specialized. Specialized crews may not be available. The number of crews depends on the time of day.

With a certain probability, a call comes in from a random address at any time. The probability of a call appearing depends on the city's area, time of day, and season. With a given probability, an incoming call has a certain priority and a call profile: general, pediatrics, psychiatry, or intensive care.

If there is an idle crew at the EMC station closest to the call site, the call is transferred to it. In this case, the appointment of the crew takes into account the call profile. If at the nearest three stations there is the necessary specialized crew, then the call is transferred to it. If the call is non-core, then the pediatric crew will go to the call only if there are no general crews at the moment.

If there is no idle crew at the EMC station closest to the call, then an unoccupied crew is selected, which will be at the closest point to the call's place. This point could be a hospital or the crew's address that has completed a previous call. In this case, if the condition of the patient served by such a crew changes to «hospitalization is required», then the call is transferred to another idle nearest ambulance crew.

When a call is received, the crew is directed to the call address. At the call's place, the crew is busy for a certain amount of time according to the given distribution function. As a result of the call service, the patient may need to be hospitalized. If such a need arises, the crew takes the patient to a suitable hospital. After that, the crew is released and goes to a new call or to the station if there are no new calls. If there is no need for hospitalization, then the crew is free immediately after the call is processed.

Patient hospitalization sites are selected not only by location but also by the profile of the call. There is a limit on the patients' number in hospitalization places. If there are no vacant places in the nearest specialized hospital, another specialized hospital is selected. If, in this case, there are no free places, then the patient is hospitalized in the nearest non-core. At the same time, if in any of the hospitalization's possible places there is an overflow of more than 10 patients, then hospitalization in this hospital is not carried out. If the overflow in all possible specialized hospitals is more than 10, then they are hospitalized in the nearest non-core, etc.

In Fig. 3, we will decompose the described algorithm and highlight the actions of the modeling's main objects, using the notation: A – the call generation subsystem, B – the subsystem for assigning the crew to the call, C – the subsystem for servicing the call by the EMC crew.

Based on the analysis, we will present a formalized model of the EMS's activities.

The resulting indicators of the EMC (Y) are determined through the indicators' values of the external flow (P_{out}), characterizing the quality of the living environment, and the indicators' values of the internal flow (P_{int}), which are formed on the quality's basis of the organization's structure and the quality of the emergency medical care process. Also, the result is influenced by random factors (P_{sl}). In this way:

$$Y = f(P_{out}, P_{int}, P_{sl}). \quad (1)$$

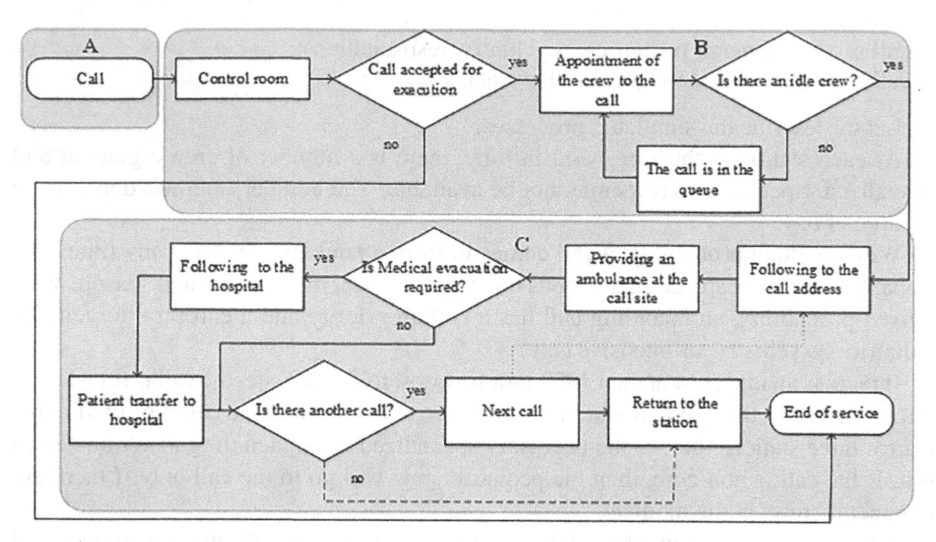

Fig. 3. Structural and functional model of the process

The goal of optimizing EMC activities is to ensure the availability and quality of emergency medical care. The model can be limited by regulatory restrictions and requirements, as well as restrictions on the financial security's amount.

The external flow is specified by the following set of incoming calls' characteristics:

$$P_{out} = \{T, Q, V, R, I, G, K, \theta\} \quad (2)$$

where T – many types of calls depending on the priority of the call (emergency, urgent); Q – multiple call profiles (general, pediatrics, psychiatry, resuscitation); V – set of incoming calls; R is a set of residential objects' geographic coordinates of a city district for each of which the relative density of calls is known; I – a set of time periods in a day, each of which has a certain density of calls (daytime – from 9:00 to 21:00 local time and night – from 21:00 to 9:00); G – many seasons of the year, which determine the seasonal course of calls; K – a set of coefficients for adjusting the mathematical expectation of the time interval between calls for the time of day (i) and the season of the call (g):

$$K = \{i, g\} \quad (3)$$

For the input stream, a set of probabilistic characteristics is set θ.

$$\Theta = \{f, M, F\}, \quad (4)$$

where f – the distribution function of the random variable «time interval between calls», M – mathematical expectation of a random variable «time interval between incoming calls», F – set of event probabilities (event's probability «call type – emergency», event's probability «call profile – pediatrics», event's probability «call profile – psychiatry», event's probability «call profile – resuscitation», the probability of the need for the patient's hospitalization).

We will assume that the mathematical expectation of the intervals between calls for a specific area, time of day, and season of the call is calculated by the formula:

$$m_l = m \cdot i_m \cdot g_n, \quad (5)$$

where m – the mathematical expectation of the interval between calls, adjusted for urgency, time of day, and season of the call, where $1 \leq l \leq |I \times G|$.

The internal flow is characterized by the quality's indicators of the organization's structure of emergency medical care and its probabilistic characteristics:

$$P_{int} = \{\Pi, C, S, \Omega\} \quad (6)$$

Π – many profiles of the EMC crews. C – many stations and substations of the EMS, for each station/substation of the EMC are known:

- geographic coordinates of the EMC station/substation location;
- number of different profiles' crews (paramedics and general doctors, as well as specialized crews: resuscitation, pediatric, psychiatric).

S – many emergency hospitals ready to receive emergency patients. For each hospital:

- geographic coordinates of the hospital location;
- number of beds.

Ω – probabilistic characteristics' set of the internal flow.

$$\Omega = \{r, R, w, W, h, H\}, \quad (7)$$

where r – the distribution function of the call reception time, R is the expectation of the call reception time, w is the distribution function of the call processing time in the control room, W is the expectation of the call processing time in the control room, h is the set of distribution functions for the time spent by the crew on the call, H is the set of mathematical expectations the time spent by the crew on calls of different types.

The researched process is characterized by *indicators of availability and quality (Y)*.

The optimized parameter that varies during iterations to obtain the normative indicators of the EMC will be the number of crews of each profile at each station/substation in each time period in the day to ensure timely availability in the provision of emergency medical care.

3 Results

Based on the proposed formalization in the AnyLogic software environment, a multi-agent simulation model was developed, which includes four objects (see Fig. 4), three of which (Calls, Dispatcher, Brigade) represent interaction at the context level: the Calls object forms calls that are processed by the "Dispatcher" object and assigned to the "Brigade" for execution. The purpose of the "Result Quality Assessment" block is collection, calculation and visualization of quality indicators. The model calculates and accumulates the following indicators of the quality of emergency medical care:

- indicators of the quality of the living environment: the total number of calls for the period, the number of emergency calls for the period;
- indicators of the quality of the structure of the organization and the process of providing emergency calls: the share of departures on emergency calls with a time of arrival in excess of the 20-min standard for the period, the average time of arrival for departures on emergency calls for the period, the average time of arrival for departures on urgent calls for the period, the average daily load brigades for the period;
- indicators of the quality of the results of the provision of emergency medical services: patient satisfaction with the service, receiving assistance in accordance with the occasion of the request, improving the state of health. Within the framework of the model, the indicators of the quality of results take values from the minimum estimate – 0 to the maximum value –1.

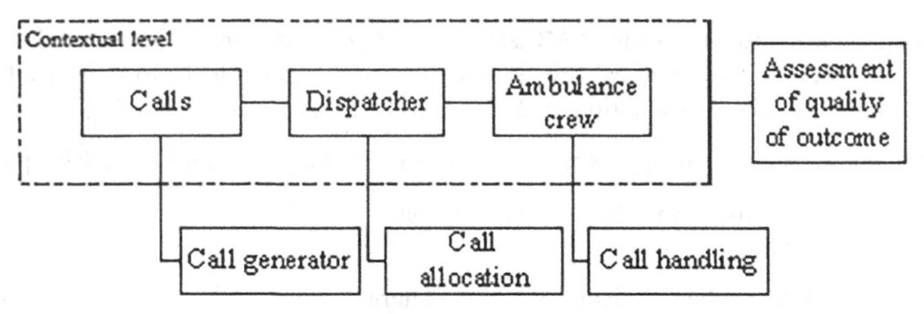

Fig. 4. Generalized structure of the simulation model

Within the context of the context level, interactions are made between those seeking medical help and the emergency medical service. The model's operation scheme is formed from the behavior of agents simulating the occurrence of calls (Call agent type), dispatchers' actions (CallDistribution agent type) and ambulance crews (Ambulance agent type), as well as an agent responsible for calculating the indicators for assessing the quality of the result (Outcome agent type). The model also contains passive agents required for modeling patients and EMS substations (Patient and Station agent types, respectively). Agents interact within the geospatial environment defined by the GIS map of Yekaterinburg, a large urban agglomeration of the Russian Federation.

The call generation submodel (Call agent type) is built using a simulation model for predicting the need for ambulance services [2] and takes into account the dynamics of changes in the total number of calls to an ambulance in the city of Yekaterinburg. In order to make the probability of a call in each point of the city correspond to the real one in modeling, to create passive agents of the patient population (type of agents Patient), the geographical coordinates of the call sites of the database of 15 342 calls collected using the epidemiological monitoring information system in 2017 were used.

The processes of call distribution and call service by the ambulance brigade are modeled using agent-based modeling tools.

In the simulation model, the experiment "variation of parameters" is implemented, designed to implement automatic repeated launch of the model with fixed values of the parameters in order to assess the mathematical expectation of the resulting indicators.

A study of the adequacy of the model was carried out, the results of which showed that the model reflects with a sufficiently high accuracy the results of the activities of the emergency medical service in Yekaterinburg.

A series of experiments has been carried out, which makes it possible to comprehensively compare the indicators of the quality of service calls with the existing level of provision of the population with institutions and ambulance stations (Experiment 1) and the level of provision planned by 2025, set by the Draft Master Plan for the Development of the Urban District – Municipal Formation "City of Yekaterinburg" for the period until 2035 (Experiment 2).

Statistically significant improvements in the following quality indicators have been proven under the conditions of Experiment 2 in comparison with the results of Experiment 1:

- an increase in the proportion of departures on emergency calls with an arrival time in excess of the 20-min standard ($p = 0.03$);
- decrease in the average daily workload of the brigade for the period ($p = 0.01$);
- patient satisfaction with service ($p = 0.05$).

Thus, the conceptual model of the quality of emergency medical care proposed in the article, which develops the content of the provisions of A. Donabedian's model and allows taking into account the influence of the quality of the living environment on the availability and quality of emergency medical care, served as the basis for the concept of modeling the quality of emergency medical care, according for evaluating the results of the EMS, sub-models should be distinguished, the totality of which makes it possible to take into account all categories of the quality of emergency medical care, taking into account the characteristics of the type of settlement. The developed multi-approach simulation model takes into account the following features of the EMC organization in a megalopolis: uneven building density, which sets a different frequency of calls for city districts; the dynamics of changes in the demand for ambulance services; actual traffic routes in the megalopolis, the number of road accidents, etc. and can be used to manage the resources of the EMC to improve the quality and availability of emergency medical care.

The results of the study can be suitable for use in the development of models of the activities of healthcare organizations, designed to develop recommendations for improving the management strategy of a medical institution.

References

1. Begicheva, S.: Fuzzy model for evaluating the quality of medical care. In: IEEE 21st Conference on Business Informatics (CBI 2019), vol. 2, pp. 5–8 (2019)
2. Begicheva, S.V.: Use of simulation modeling to select input variables in predictive models for the demand for emergency medical services in Russia. In: 2020 IEEE 22nd Conference on Business Informatics (CBI 2020), pp. 101–105 (2020)
3. Cabrera, E., Taboada, M., Iglesias, M., Epelde, F., Luque, E.: Simulation optimization for healthcare emergency departments. In: 12th Annual International Conference on Computational Science, ICCS 2012, pp. 1464–1473. Elsevier, Omaha (2012)

4. Carayon, P., et al.: Work system design for patient safety: the SEIPS model. Qual. Saf. Health Care Global J. Health Sci. **6**(4) (2014)
5. Coyle, Y.M., Battles, J.B.: Using antecedents of medical care to develop valid quality of care measures. Int. J. Qual. Health Care: J. Int. Soc. Qual. Health Care/ISQua **11**(1), 5–12 (1999)
6. Decree of the Government of the Russian Federation of December 26, 2017 N 1640 "On approval of the state program of the Russian Federation" Development of health care "(with amendments and additions)
7. Donabedian, A.: Quality assessment and assurance: unity of purpose, diversity of means. Inquiry **25**(1), 173–192 (1988)
8. Federal Law of 21.11.2011 N 323-FZ (as amended on 29.05.2019) "On the basics of protecting the health of citizens in the Russian Federation", Art. 2
9. Guidelines for emergency medical care [Text]/Ministry of Health and Social Development of the Russian Federation, ASMOK. GEOTAR-Media, Moscow (2007)
10. Korobkova, O.K.: Problems of improving the provision of medical services in rural areas of the regions of the Russian Federation. Actual Probl. Econ. Law **1**(33), 179–186 (2015)
11. Laker, L.F., et al.: Understanding emergency care delivery through computer simulation modeling. Acad. Emerg. Med. **25**, 116–127 (2018)
12. Mitchell, P.H., Ferketich, S., Jennings, B.M.: Quality health outcomes model. Am. Acad. Nurs. Expert Panel Qual. Health Care **30**(1), 43–46 (1998)
13. National Highway Traffic Safety Administration: Configurations of EMS Systems: a Pilot Study. Ann Emerge Med **52**, 453 (2008)
14. Nonukova, I.V.: Organization of medical care in conditions of inaccessibility of places of residence and low population density (on the example of the Altai Republic). Novosibirsk – LLC "Alfa-Resource" (2012)
15. Nuckols, T.K., Escarce, J.J., Asch, S.M.: The effects of quality of care on costs: a conceptual framework. Milbank Quart. **91**(2), 316–353 (2013)
16. Order of the Ministry of Regional Development of Russia dated 09.09.2013 N 371 "On approval of the methodology for assessing the quality of the urban living environment"
17. Sharabchiev, Y., Dudina, T.V.: Accessibility and quality of medical care: components of success. Int. Rev.: Clin. Pract. Health **4**, 16–34 (2013)

Fuzzy Risk States Assessment Using Markov Chains

Mikhail Matveev and Vladislav Korotkov(✉)

Voronezh State University, Voronezh, Russia
chasecrunk@gmail.com

Abstract. The paper is devoted to automated risk assessment for decision making in organizational systems with credit risk as an example. The classical approach involves calculating the borrower's credit rating using some information model trained on a statistical ensemble of its indicator values. However, some of the indicators change over time, and their dynamics is not directly taken into account, which reduces the reliability of the risk assessment. To consider the dynamics of indicators, it is proposed to use the Markov chain working with fuzzy states of an indicator specified on the linguistic scale of these states. Fuzzy states are not observable, so an algorithm for calculating the elements of transition probability matrix based on a time series of indicator values is proposed. The Markov chain allows you to obtain stationary probabilities of fuzzy states, which are converted into average values of the indicator. Numerical examples show that in the case of a stationary time series, the classical mean of an indicator equals to an estimate obtained by the proposed method. If the dynamics of an indicator is increasing, then the classical mean gives an underestimation of its true value. If the dynamics is decreasing, then the classical mean gives an overestimated value.

Keywords: risk assessment · credit risk · markov chain · fuzzy states · fuzzy time series

1 Introduction and Problem Definition

Assessment of the state of a dynamic system is an important decision-making task. In a number of applications, the set of states of the object under consideration is specified on a continuous scale. In this case, the decision maker has problems assessing the state. An example is a credit score, the calculation of which is based on the borrower's credit history (an example is shown in Table 1). Credit scoring is an assessment of the risk of a lender issuing a loan to a potential borrower: an individual, business, company or government. The specific algorithm for calculating a credit score is always a state or commercial secret. However, historical data on previous loans of the borrower, the level of his income and a number of other indicators that determine the risk of the lender are usually used for the calculation. Historical data always includes the dynamics of change in the values of a set of features over a certain period of time. In this case, it is called behavioral scoring [1]. Historical data can be formalized in the form of multidimensional

V. Taratukhin et al. (Eds.): ICID 2022, CCIS 1767, pp. 15–26, 2023.
https://doi.org/10.1007/978-3-031-32092-7_2

time series. But most of the well-known approaches to solving the problem of credit scoring, such as logistic regression, decision trees, multilayer perceptrons, etc., use data in the form of statistical ensembles and do not take into account the dynamics of change, which can lead to a significant risk assessment bias. Recurrent neural networks (RNNs) take into account the dynamics, but their implementation is not trivial and their training requires a lot of statistical data in the form of long time series. Relatively simple dynamic models like Markov chains (MC), which are used in behavioral scoring [1–3], can serve as an alternative to RNNs. An important feature of the MC is the ability not only to predict the transition states of a dynamic object, but also to determine stationary states that are important for assessing the risk status.

The risk scale is often given as probabilities in the range [0; 1] and is divided into several classes (statuses). An example of such a ranking is shown in Table 1.

Table 1. Example of ranking credit risk values

Probability of issuing a loan	Status	Status interpretation
0,951–1	High	The probability of loan rejection is extremely small
0,896–0,950	Good	Good chances of getting a loan
0,761–0,895	Average	Getting a loan is possible, but not guaranteed
0,596–0,760	Bad	The probability of obtaining a loan is extremely small
0–0,595	Very bad	Getting a loan is almost impossible

It is difficult to make reasonable decisions on the boundaries of the presented equivalence classes of risk probability values. It is more appropriate to consider tolerance classes. The statements in the right column of Table 1 therefore can be interpreted as term sets of a linguistic variable $\widetilde{KR}$ with the name «Probability of getting a loan». We will consider the set of term sets as a set of fuzzy credit risk states with the names presented in the middle column of Table 1. In this case, a fuzzy state Markov chain model can be used to obtain risk estimates, which is more flexible and realistic than the classical Markov chain model based on crisp states. Methods for constructing a Markov chain for fuzzy states is an actual topic of many studies. For example, Zadeh [4] introduced fuzzy sets for the mathematical description of fuzzy states. Kruce et al. [5] presented the fuzzy Markov chain as a classical Markov chain based on fuzzy transition probabilities. The main problem is calculating these probabilities. Pardo and Fuente [6] showed an approach to calculating transition probabilities based on the concept of the conditional probability of a fuzzy state. Dao Xuan Ky and Luc Tri Tuyen [7] considered the construction of a transition matrix based on the results of observing a fuzzy time series. When building a Markov chain for fuzzy states of credit risk, there is an observable dynamics of historical data in the form of time series, while the fuzzy states themselves are not observable.

The objective of the study is to build a Markov chain model for fuzzy states, automatically calculate the risk status and study the influence of considering the dynamics on the accuracy of the risk status assessment.

2 Fuzzy Time Series and Markov Chain

This section presents brief information about fuzzy time series and Markov chains, based on the works [4–6, 8, 9].

Definition 1. A fuzzy set is a subset $\tilde{A}$ of the universal discrete set X given by the following expression.

$$\tilde{A} = \{x_1|\mu(x_1); x_2|\mu(x_2); \ldots; x_n|\mu(x_n)\}, \quad (1)$$

where $x_i \in X$ – historical data values, $\mu(x_i) \in [0; 1]$ – membership function values for element x_i and set U.

Definition 2. The fuzzy state of credit risk $\tilde{A}_i$ is the fuzzy set defined on the scale of the observed risk factor x with a normalized triangular membership function.

$$\mu_{\tilde{A}_i}(x) = \begin{cases} 0, x \leq a_i; \\ \frac{x-a_i}{c_i-a_i}, a_i < x \leq c_i; \\ \frac{b_i-x}{b_i-c_i}, c_i < x < b_i; \\ 0, x \geq b_i. \end{cases} \quad (2)$$

Definition 3. Linguistic variable $\widetilde{KR}$ called «Credit risk status» is a variable with a given finite term-set of states $\{\tilde{A}_1; \ldots; \tilde{A}_s\}$, the semantics of which is determined using a set of membership functions $\{\mu_{\tilde{A}_i}(x)\}$, where x is the value of a dynamically changing risk factor. The term-set of states and the corresponding membership functions are set by an expert on the scale of the corresponding risk factor.

Definition 4. A linguistic scale of fuzzy risk states is an ordered set of fuzzy states $\tilde{A}_i$ such that if $i < j$, then $\tilde{A}_i$ precedes $\tilde{A}_j$ ($\tilde{A}_i \prec \tilde{A}_j$). A possible view of such a linguistic scale is shown in Fig. 1.

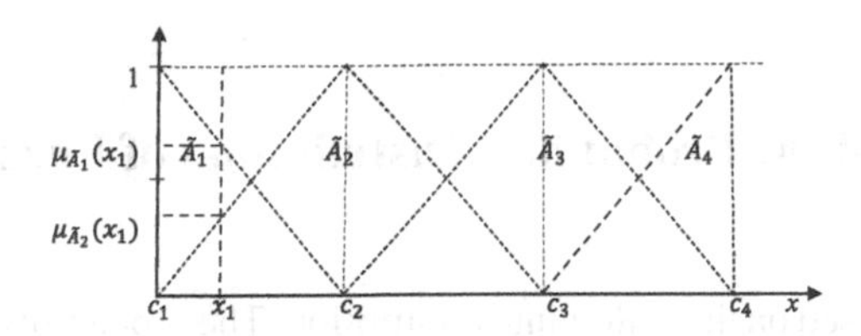

Fig. 1. Linguistic scale example.

To assess fuzzy risk states, the levels of the time series of historical data generated from the elements of the set X is used. The assessment is based on the correspondence between the level of the series and the fuzzy state. An important condition that must be observed when forming a language scale is $\sum_{i=1}^{S} \mu_{\tilde{A}_i}(x) = 1$.

Definition 5. Fuzzy correspondence $C : X \to \left\{\tilde{A}_i | \mu_{\tilde{A}_i}(x)\right\}$ is a correspondence between the level of the time series, $x \in X$, and the fuzzy state $\tilde{A}_i$. The correspondence value is determined at time t by the value of the membership function $\mu_{\tilde{A}_i}(x^t)$. In the general case, the fuzzy correspondence is ambiguous, one value x^t can correspond to several fuzzy states $\tilde{A}_i$ with their own values of membership functions. An example of ambiguity is shown in Fig. 1.

Definition 6. A fuzzy time series $\tilde{A}(t)$ is a sequence of randomly changing fuzzy risk states ordered by time t – $(\tilde{A}_i^0; \tilde{A}_i^1; \ldots; \tilde{A}_i^t \ldots)$, where $\tilde{A}_i^t \in \{\tilde{A}_1; \ldots; \tilde{A}_s\}$.

Let a random sequence $S_n; n \in N$ be given, in which S_n takes a countable number of values $I = \{1, 2, \ldots, m\}$.

Definition 7. A random sequence of system states is called a Markov chain if and only if for all $(s_0; s_1; \ldots; s_n) \in I$ the following condition is satisfied:

$$P(S_n = s_n | S_0 = s_0; \ldots; S_{n-1} = s_{n-1}) = P(S_n = s_n | S_{n-1} = s_{n-1}). \quad (3)$$

Let the state vector $(s_1; s_1; \ldots; s_m)$ at a given time t be characterized by the corresponding probability distribution vector of the system states at this moment – $p^t = (p_1^t; \ldots; p_m^t)$.

The recursion of the vector p^t is defined by the following expression

$$p^t = M \cdot p^{t-1}, \quad (4)$$

where $M = (p_{ij})$ is the probability matrix of system transitions from state i to state j; $\forall i (\sum_j p_{ij} = 1)$. If the matrix M does not change with time, then the Markov chain is called homogeneous. The probability distribution vector p^t describes changes in distributions during transient processes.

Definition 8. A stationary distribution p^s is defined as $\lim_{t \to \infty} p^t = p^s$, where $p^s = Mp^s$. The stationary distribution is interpreted as a status of the system.

3 Calculation of the Probability Distribution of Fuzzy Credit Risk States

Further material is based on the following assumption. The probabilities of credit risk are determined on a certain time interval $[\tau_0; \tau_k]$ by a set of dynamically changing factors from the set of historical data. Moreover, each factor in one way or another affects the probability of risk. Accounting for the influence of each of them should be carried out by the corresponding Markov chain. The resulting local probability distributions of risk states will be further aggregated taking into account the degree of influence of each factor and their possible interaction.

This section provides a methodology for calculating the probability distribution of credit risk for a single dynamically changing factor.

Let us consider the main stages of the proposed modeling technique. To construct a fuzzy time series model, it is necessary to specify a linguistic scale in the form of an ordered set of terms $(\tilde{A}_1; \tilde{A}_2;\ldots; \tilde{A}_s)$. This is done as follows [10]. A continuous segment $[a; b] \subseteq X$ is chosen in such a way that this segment includes all possible values of the levels x^t of the time series. A fuzzy partition of $[a; b]$ is made, i.e., a term-set of states $(\tilde{A}_1; \tilde{A}_2; \ldots; \tilde{A}_s)$ is specified, corresponding to the following assumptions:

- membership functions $\mu_{\tilde{x}}(x)$ are normalized triangular functions with modes equal to one and zero values at the ends of the support;
- to ensure uniform coverage of $[a; b]$, the supports of adjacent membership functions $\mu_{\tilde{x}_i}(x)$ are chosen and placed on the set $[a; b]$ so that they intersect and the equality $\mu_{\tilde{x}_{i-1}}(x^\star) = \mu_{\tilde{x}_i}(x^\star) = 0,5$ is satisfied at the intersection points $x^\star$.

Let $x^t = (x^0; x^1; \ldots; x^n)$ be a time series of risk factor values, the levels of which are given by discrete values or intervals (in the case of continuous x) on $[a; b]$. Let us also define a linguistic scale containing the necessary and sufficient number of terms to describe fuzzy states.

The fuzzy time series of states is formed on the basis of the introduced concept of fuzzy correspondence $C : X \to \{\tilde{A}_i | \mu_{\tilde{x}_i}(x)\}$, which is set at each time t. We will successively substitute the values x^t into all membership functions $\mu_{\tilde{x}_i}(x^t)$; $i = 1, \ldots, s$; $t = 0, \ldots, n$ and mark those states whose values are greater than zero. The marked membership functions, due to the ambiguity of the correspondence C, will determine the fuzzy vector of fuzzy states at time t:

$$\tilde{A}^t = (\tilde{A}_i^t | \mu_{\tilde{A}_i}(x^t) \neq 0, i = 1; \ldots; k). \tag{5}$$

Thus, the change over time in the values of the risk factor x determines a fuzzy time series of states $(\tilde{A}^1; \ldots; \tilde{A}^t; \ldots)$. If we assume that the time series of risk factor values has the Markov property, then the fuzzy time series of states, due to the adopted algorithm for its construction, will also correspond to the Markov property. In this case, following the classical methods, in particular, works [11, 12], it can be argued that the previous state $\tilde{A}_i^{t-1} \in \tilde{A}^{t-1}$ with a certain probability will be followed by the state $\tilde{A}_j^t \in \tilde{A}^t$, and, therefore, on the set of states, one can introduce a fuzzy relation $R_{ij}(t-1, t) = (\tilde{A}_i^{t-1}, \tilde{A}_j^t)$, which can be formally represented using fuzzy implications $A_i^{t-1} \to \tilde{A}_j^t$. We will consider such a fuzzy relation as a fuzzy state change event.

At time t, several fuzzy events appear at once, determined by the direct product of fuzzy vectors $R^{t-1,t} = \tilde{X}^{t-1} \times \tilde{X}^t$. Then each binary relation defined by such a direct product can be written as $\tilde{A}_i^{t-1} \to \tilde{A}_j^t | \mu_{ij}$, where the value of the membership function is given by the following expression:

$$\mu_{ij} = \min\left(\mu_{\tilde{A}i};\ \mu_{\tilde{A}j}\right). \tag{6}$$

The observed features of the state x are usually considered as random variables, due to the randomness of the factors that form them and (or) the presence of observation errors. Therefore, the event of taking some particular value x^t is random and is characterized by a probability $p(x^t) \in [0; 1]$. Since the fuzzy state change event is defined with x^t, it should also be considered as a random event. The combination of two distinct types of uncertainty: fuzziness and randomness is found in many studies, for example [13–16], and requires the introduction of the concept of a fuzzy probability space.

According to [15], the fuzzy probability space is defined by the triple (Ω, $\sigma(\Omega)$, $P(\tilde{A})$), where Ω – the set of fuzzy random events; $\sigma(\Omega) \subset 2^{\Omega}$ – the sigma-algebra on the set Ω; $P(\tilde{A})$ – the probability of a fuzzy random event $\tilde{A} \in 2^{\Omega}$. The fuzzy probability space is a generalization of the classical probability space.

Definition 9. The random event of the system transition from the fuzzy state $\tilde{A}_i$ to the fuzzy state $\tilde{A}_j$ at time t with the degree of confidence μ_{ij} is called a fuzzy elementary event $\tilde{A}^{э}_{ij} = (\tilde{A}^{t-1}_i \to \tilde{A}^{t}_j)|\mu_{ij}$. The random nature of the event $\tilde{A}^{э}_{ij}$ is due to the random sequence of indicator values; the fuzzy character of $\tilde{A}^{э}_{ij}$ is due to the fuzzy implication – $(\tilde{A}^{t-1}_i \to \tilde{A}^{t}_j)|\mu_{ij}$.

Note that fuzzy correspondences C at various adjacent times can generate homogeneous (in the sense of the invariance of the pair ij) implications with different values of the membership function:

$$\tilde{A}_{ij} = \{(\tilde{A}^{t_1-1}_i \to \tilde{A}^{t_1}_j)|\mu^1_{ij}; \ \ldots\ldots; (\tilde{A}^{t_k-1}_i \to \tilde{A}^{t_k}_j)|\mu^k_{ij}\}. \quad (7)$$

Definition 10. A fuzzy event $\tilde{A}_{ij}$ is a composite event that includes all homogeneous transitions $\tilde{A}_i \to \tilde{A}_j$ with different values of membership functions – $U_k(\tilde{A}_i \to \tilde{A}_j)|\mu^{kij}_{ij}$.

L. Zadeh [16] defined the probability of a fuzzy event $\tilde{A}$ relative to the distribution $P(\tilde{A})$ as the following expression:

$$P(\tilde{A}_{ij}) = \textstyle\sum_k p(\tilde{A}^{эk}_{ij})\mu^k_{ij}, \quad (8)$$

which can be interpreted as a weighted average of the probabilities of fuzzy elementary random events, where the weights are the values of membership functions for transitions over a given time series interval.

An estimate of the probability distribution $P(\tilde{A}_{ij}) = \sum_k p(\tilde{A}^{эk}_{ij})\mu^k_{ij}$, can be obtained based on the concept of statistical probability. To do this, we represent all possible fuzzy random events on the set of fuzzy states ($\tilde{A}_1; \tilde{A}_2; \ldots; \tilde{A}_s$) as a matrix of dimension $s \times s$:

$$\begin{pmatrix} U_{k11}(\tilde{A}_1 \to \tilde{A}_1)\big|\mu^{k11}_{11} & \ldots & U_{k1s}(\tilde{A}_1 \to \tilde{A}_s)\big|\mu^{k1s}_{1s} \\ \ldots & \ldots & \ldots \\ U_{ks1}(\tilde{A}_s \to \tilde{A}_1)\big|\mu^{ks1}_{s1} & \ldots & U_{kss}(\tilde{A}_s \to \tilde{A}_s)\big|\mu^{kss}_{ss} \end{pmatrix}. \quad (9)$$

The elements of matrix (9) are the unions of all homogeneous fuzzy events obtained as a result of observations. Let us denote the maximum value of the index k_{ij} as K_{ij} – the number of homogeneous events in the union ij. It is easy to see that each row of matrix (9) contains a complete group of events. The number of elementary fuzzy events in each row of matrix (9) is equal to $K_i = \sum_{j=1}^{s} K_{ij}$ for all i. Let's select elementary events with identical values of μ^k_{ij} from each set of homogeneous events. Let's also denote the number of such values as K_{ij}^{μ}. Then the frequency or statistical probability of an elementary event $p(\tilde{A}_{ij}^{эk})$ in the complete group of events is calculated according to:

$$w(\tilde{A}_{ij}^{эk}) = p(\tilde{A}_{ij}^{эk}) = \frac{K_{ij}^{\mu}}{K_i}. \tag{10}$$

Expression (10) allows calculating the probabilities (8) for the matrix elements (9). The probability matrix corresponding to matrix (9) will look like:

$$\begin{pmatrix} P\left(\tilde{A}_{11}\right) \ldots P\left(\tilde{A}_{1s}\right) \\ \ldots \quad \ldots \quad \ldots \\ P\left(\tilde{A}_{s1}\right) \ldots P\left(\tilde{A}_{ss}\right) \end{pmatrix}. \tag{11}$$

4 Markov Chain of Fuzzy Random Events and Determination of Risk Status

The probabilities of fuzzy random events $P(\tilde{A}_{ij})$ could be considered as elements of the stochastic transition matrix M in expression (4), but these probabilities, generally speaking, do not satisfy the mandatory condition:

$$\forall i (\sum_j P_{ij} = 1). \tag{12}$$

Condition (12) will be satisfied after the following normalization P_{ij}:

$$\forall i \left(P_{ij} = \frac{P(\tilde{A}_{ij})}{\sum_j P(\tilde{A}_{ij})} \right). \tag{13}$$

Now the obtained values p_{ij} can be considered as elements of the stochastic transition matrix in (4) and interpreted as the probabilities of random fuzzy events – the transition from the fuzzy state $\tilde{A}_i$ to the fuzzy state $\tilde{A}_j$. The discrete model of the Markov chain of random fuzzy states is no different from the classical notation:

$$\begin{pmatrix} p_1^{t+1} \\ \ldots \\ p_s^{t+1} \end{pmatrix} = \begin{pmatrix} p_{11} \cdots p_{s1} \\ \ldots\ldots\ldots \\ p_{1s} \cdots p_{ss} \end{pmatrix} \cdot \begin{pmatrix} p_1^t \\ \ldots \\ p_s^t \end{pmatrix} \tag{14}$$

with initial condition $\left(p_1^0, \ldots, p_s^0\right)^T$.

To determine the status of the system, i.e., to determine the probabilities of its stationary states $p^s = (p_1^s; p_2^s; \ldots; p_s^s)$, it is necessary to solve system (14) with the stationarity condition – $p^{t+1} = p^t$. To obtain nontrivial solutions, system (14) is usually rewritten with one equation replaced by a normalization equation, $p_1^s + p_2^s + \ldots + p_s^s = 1$.

5 Numerical Experiments

Numerical experiments are needed to test the claimed advantages of the proposed method for constructing a Markov chain with fuzzy risk states and to demonstrate automatic assessment of the risk state when processing a hypothetical borrower's time series of data.

In order to explore the advantages of the proposed algorithms, a special technique was developed that makes the difference in the static and dynamic approaches clearer. Taking into account the assumptions of Sect. 3, we will consider a scalar hypothetical risk factor represented by a time series on a ten-point scale:

$$x = \left(x^1; x^2; \ldots; x^{12}\right) = (5; 6; 6; 5; 6; 6; 5; 6; 6; 5; 6; 6).$$

A statistically significant regression dependence of series levels on time is described by the equation $x = at + b$, where $a = 0$, $b = 5,667$. I.e., the time series is stationary with an average value of 5.667.

Let a linguistic variable with the name «Probability of obtaining a loan», with the corresponding fuzzy statuses: «$\tilde{A}_1$ – high»; «$\tilde{A}_2$ – good»; «$\tilde{A}_3$ – medium»; «$\tilde{A}_4$ – low», be represented on the linguistic scale shown in Fig. 2.

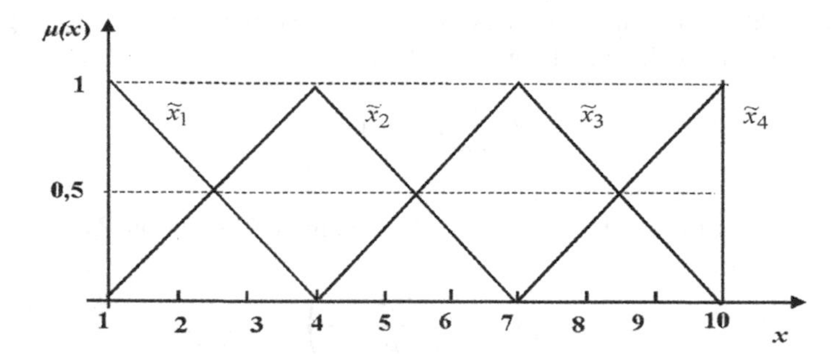

Fig. 2. Linguistic scale for probability of obtaining a loan

Applying the proposed method described in Sects. 2 and 3, we obtain matrix of transition probabilities of fuzzy risk states:

$$\boldsymbol{M} = \begin{pmatrix} 0 & 0 & 0 & 0 \\ 0 & 0{,}423 & 0{,}577 & 0 \\ 0 & 0{,}483 & 0{,}517 & 0 \\ 0 & 0 & 0 & 0 \end{pmatrix}.$$

The corresponding system of Eqs. (14), taking into account the normalization equation, will look like:

$$\begin{cases} (0{,}423 - 1)p_2^c + 0{,}483p_3^c = 0, \\ p_2^c + p_3^c = 1. \end{cases}$$

The solution of this system of equations gives us the following vector of stationary states $p^s = \{0;\ 0,456;\ 0,544;\ 0\}$.

To check the adequacy of the proposed methodology, let's digitize the risk statuses, for example, «5 – high»; «4 – good»; «3 – medium»; «2 – low». Such a digitization will allow us to calculate the arithmetic mean of the statuses, the probabilities of which are estimated using the Markov chain

$$\overline{x}^{MC} = 0 \cdot 2 + 0{,}456 \cdot 3 + 0{,}544 \cdot 4 + 0 \cdot 5 \approx 3{,}544.$$

Digitization allows you to more accurately assess the status. In this case, the status is between «good» and «average», a little closer to good.

It is reasonable to expect that the average value of the stationary time series should be equal to the mathematical expectation of the stationary status values obtained using the Markov chain. To compare the average values of the time series and the average values of fuzzy states, it is necessary to establish a correspondence between the measuring scales of the average values. The linear correspondence of the scales is easy to build on the basis of the correspondences between the beginning and the end of each scale. In this case, the linear correspondence of the scale [1; 10] and the scale [2; 5] will look like:

$$\overline{x}^{[2;5]} = \frac{1}{3} \cdot \overline{x}^{[1;10]} + \frac{5}{3}. \tag{15}$$

The value of the average time series of local estimates given by correspondence (15) is:

$$\overline{x}^{[2;5]} = \frac{1}{3} \cdot 5{,}667 + \frac{5}{3} \approx 3{,}556.$$

As can be seen, the average value of risk factor time series and the average value of the digitized stationary states are almost equal, as expected. This can be considered as an experimental confirmation of the adequacy of the proposed method.

Let us consider the non-stationary time series of risk factor values. We will consider two non-stationary series, ascending and descending, respectively (Fig. 3):

$$x_a = (2;\ 3;\ 3;\ 4;\ 5;\ 5;\ 6;\ 6;\ 7;\ 8;\ 9;\ 10)$$

$$x_d = (10;\ 9;\ 8;\ 7;\ 6;\ 6;\ 5;\ 5;\ 4;\ 3;\ 3;\ 2)$$

Note that the average values of the presented series are 5.667, i.e., they are equal to the average of the stationary series. Figure 3 shows the mean value as a dotted line.

The corresponding matrices of transition probabilities of fuzzy risk states are:

$$\boldsymbol{M}_{\mathrm{a}} = \begin{pmatrix} 0{,}33 & 0{,}667 & 0 & 0 \\ 0{,}105 & 0{,}579 & 0{,}316 & 0 \\ 0 & 0{,}188 & 0{,}562 & 0{,}25 \\ 0 & 0 & 0{,}25 & 0{,}75 \end{pmatrix},$$

$$\boldsymbol{M}_{\mathrm{d}} = \begin{pmatrix} 0{,}5 & 0{,}5 & 0 & 0 \\ 0{,}222 & 0{,}611 & 0{,}167 & 0 \\ 0 & 0{,}375 & 0{,}562 & 0{,}062 \\ 0 & 0 & 0{,}571 & 0{,}429 \end{pmatrix}.$$

The corresponding systems of equations of type (14) will take the form:

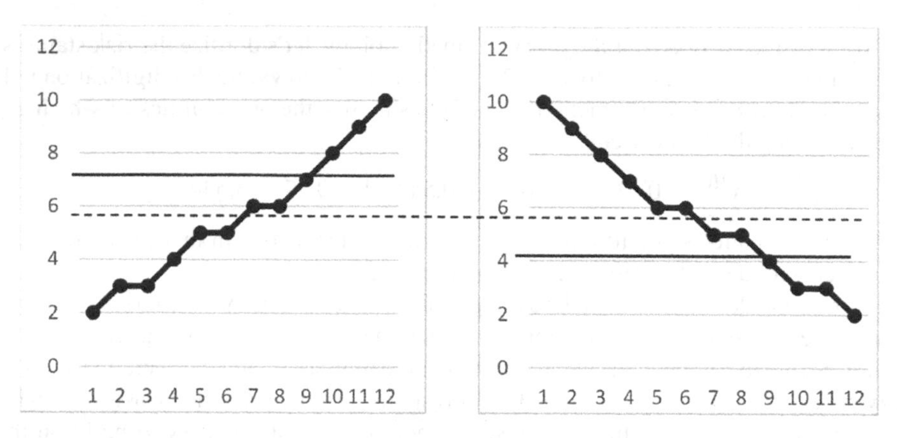

Fig. 3. Ascending (left) and descending (right) time series

- for the ascending series

$$\begin{cases} (0{,}33 - 1)p_1^c + 0{,}105p_2^c = 0, \\ 0{,}667p_1^c + (0{,}579 - 1)p_2^c + 0{,}188p_3^c = 0, \\ 0{,}316p_2^c + (0{,}562 - 1)p_3^c + 0{,}25p_4^c = 0, \\ p_1^c + p_2^c + p_3^c + p_4^c = 1. \end{cases}$$

- for the descending series

$$\begin{cases} (0{,}5 - 1)p_1^c + 0{,}222p_2^c = 0, \\ 0{,}5p_1^c + (0{,}611 - 1)p_2^c + 0{,}375p_3^c = 0, \\ 0{,}167p_2^c + (0{,}562 - 1)p_3^c + 0{,}571p_4^c = 0, \\ p_1^c + p_2^c + p_3^c + p_4^c = 1. \end{cases}$$

The solution of these equations gives us a stationary state probability distribution:

- for the ascending series:

$$p^c = \{0{,}035; 0{,}221; 0{,}372; 0{,}372\};$$

- for the descending series:

$$p^c = \{0{,}229; 0{,}516; 0{,}229; 0{,}025\}.$$

Mean of stationary states:

- for the ascending series:

$$\overline{x}^{MC} = 0{,}035 \cdot 2 + 0{,}221 \cdot 3 + 0{,}372 \cdot 4 + 0{,}372 \cdot 5 = 4{,}081;$$

In terms of the scale [1: 10] $\overline{x}^{[1;10]} = 7,244$.

- for the descending series:

$$\overline{x}^{MC} = 0{,}231 \cdot 2 + 0{,}519 \cdot 3 + 0{,}231 \cdot 4 + 0{,}019 \cdot 5 = 3{,}05.$$

In terms of the scale [1: 10] $\overline{x}^{[1;10]} = 4,151$.

Therefore, ignoring the dynamics of the risk factor leads to the establishment of an incorrect correspondence between the factor values and the probability values.

The purpose of demonstrating automatic diagnostics of the risk status in terms of states is to indicate the higher information content of the proposed approach to support decision-making on issuing a loan. Let the company-borrower have long been credited in some bank. The credit history of the borrowing company is represented by a time series of values of the loan service quality indicator calculated by the bank on a continuous scale similar to the scale in Table 1. The credit history contains information on the basis of which the bank decides on the possibility of issuing a loan and the amount of the loan rate for the borrower. A rational approach to decision-making should assume the presence of some threshold value of the time-aggregated indicator of quality of service and the allocation of classes or statuses with a fixed credit rate. The proposed approach allows, based on the analysis of credit history, to estimate the probabilities p_i^c, $i = 1, \ldots, S$ of the borrower's status matching the given statuses, for example, such as in Table 1, where $S = 5$. If the bank has determined a fixed credit rate k_i for each status, then the mathematical expectation of risk coverage can be calculated as

$$\overline{k} = \sum\nolimits_{i=1}^{S} p_i^c k_i. \tag{16}$$

Thus, having set the desired amount of coverage, the bank can vary the values of the credit rate, so as to achieve a given amount of coverage for a borrower with a specific credit history.

Let the credit history of the borrowing company be represented by the following time series of values of the quality of service for previous loans:

$$(0{,}953;\ 0{,}875;\ 0{,}920;\ 0{,}870;\ 0{,}810;\ 0{,}750;\ 0{,}960;\ 0{,}895;\ 0{,}951).$$

Stationary states (statuses) of the borrowing company according to the proposed algorithms are:

$$p^c = \{0{,}130;\ 0{,}462;\ 0{,}333;\ 0{,}074;\ 0\}.$$

Then, for the given credit rates $k = \{0,05; 0,08; 0,13; 0,20; 0,30\}$, the calculated value of coverage will be $0,13{\cdot}0,05{+}0,462{\cdot}0,08{+}0,333{\cdot}0,13{+}0,074{\cdot}0,20{+}0{\cdot}0,30 = 0,102 = 10,2\%$. Let the risk coverage defined by the bank be 12%. It is clear that the distribution of credit rates must be changed. One of the possible distributions is $k = \{0,06; 0,11; 0,15; 0,20; 0,30\}$.

6 Conclusion

As a result of the research, a method was developed for constructing a homogeneous Markov chain for a system with fuzzy states, based on the processing of time series of risk factors.

Unlike well-known fuzzy Markov chains, the transition matrix is not considered as a fuzzy relation, but remains a regular stochastic matrix. This approach makes it possible to obtain stationary risk states that characterize the risk status at some given time interval. This information can be used in a decision support system to manage lending parameters.

The key feature of the resulting Markov chain is an ability to consider the dynamics of risk factor states change when calculating the average values of risk status under conditions of fuzzy status assignment.

Numerical experiments with the constructed Markov chain of fuzzy states are presented, which confirm its adequacy and the possibility of taking into account the system dynamics. It's also demonstrated how the proposed approach can be applied to determine lending rates for a borrower with a specific credit history.

References

1. Behavioral Scoring: https://www.openriskmanual.org/wiki/Behavioral_Scoring. Last accessed 8 Oct 2022
2. Kozminyh, O.V.: Markov chains as a tool for risk assessment of insurance companies arising during the transfer of functions to outsourcing. Rossiyskoe predprinimatelstvo **18**(17), 2492–2504 (2017)
3. Madden, M.G.: Evaluation of the performance of the Markov blanket Bayesian classifier algorithm. Technical Report NUIG-IT-011002, Department of Information Technology, National University of Ireland, Galway (2002)
4. Zadeh, L.A.: Fuzzy sets. Inf. Control **8**(3), 338–353 (1965)
5. Kruce, R., Buck-Emden, R., Cordes, R.: Processor power considerations: an application of fuzzy markov chains. Fuzzy Sets Syst. **21**(3), 289–299 (1987)
6. Pardo, M.J., Fuente, D.: Fuzzy markovian decision processes: application to queueing systems. Comput. Math. Appl. **60**(9), 2526–2535 (2010)
7. Ky, D.X., Tuyen, L.T.: A markov-fuzzy combination model for stock market forecasting. Int. J. Appl. Math. Stat. **55**(3), 109–121 (2016)
8. Baldwin, J.F., Martin, T.P., Rossiter, J.M.: Time series modeling and prediction using fuzzy trend information. In: Proceedings of the International Conference SC Information/Intelligent System, pp. 499–502 (1998)
9. Batyrshin, I., et al.: Moving approximation transform and local trend associations in time series data bases. In: Perception-based Data Mining and Decision Making in Economics and Finance. Studies in Computational Intelligence, vol. 36, pp. 55–83. Springer, Heidelberg (2007)
10. Perfilieva, I.: Fuzzy transforms: theory and applications. Fuzzy Sets Syst. **157**(8), 993–1023 (2006)
11. Song, Q., Chissom, B.: Forecasting enrollments with fuzzy time series – Part I. Fuzzy Sets Syst. **54**(1), 1–9 (1993)
12. Song, Q., Chissom, B.: Forecasting enrollments with fuzzy time series – Part II. Fuzzy Sets Syst. **62**(1), 1–8 (1994)
13. Yager, R.: A note on probabilities of fuzzy events. Inf. Sci. **18**(2), 113–129 (1979)
14. Meyers, R.A. (ed.): Encyclopedia of Physical Science and Technology, 3rd edn. Academic Press, San Diego (2001)
15. Guixiang, W., Yifeng, X., Sen, Q.: Basic Fuzzy Event Space and Probability Distribution of Probability Fuzzy Space. Mathematics **7**(6), 542 (2019)
16. Zadeh, L.A.: Probability measures and fuzzy events. J. Math. Anal. Appl. **23**(2), 421–427 (1968)

Principles of Multiple Alternatives in Complex Control Systems with a Reference Model

Semyon Podvalny and Eugeny Vasiljev(✉)

Voronezh State Technical University, 394006 Voronezh, Russia
vgtu-aits@yandex.ru

Abstract. The work is devoted to the control problem of large systems, the parameters of which change in time. It is noted that with the increase in the complexity of such objects, the reliability of the corresponding control systems decreases. In this regard, the task of developing new approaches to the control of large systems remains relevant. As such an approach, it is proposed to use the evolutionary concept of multiple alternatives, which involves the reproduction of processes occurring in large ecosystems. In particular, this concept includes several principles: the principle of multi-level, the principle of diversity and separation of functions, as well as the principle of modularity. An example of applying these principles to control a complex dynamic system with a reference model is considered. The structure of a multilevel control system has been developed, which contains two channels: the channel of hierarchically connected elements of the object and the channel of the corresponding cascades of the reference model. It is shown that the possibility of such a multilevel control is based on the condition of representing a complex object in the form of elements of a low dynamic order. When this condition is met, the contradiction between the original object complexity and the control system stability is removed. A comparative analysis of the proposed control structure with known signal adaptation schemes has been given. The results of numerical simulation are presented, confirming the effectiveness of the proposed approach to the control of complex non-stationary systems.

Keywords: Complex Control Systems · Evolutionary Approach · Principles of Multiple Alternatives · Systems with a Reference Model

1 Introduction

The control of complex systems of any nature inevitably encounters a large dimension of its vector coordinates and non-stationarity of the parameters. These features of complex objects lead to the fact that the corresponding complexity of the control system for such objects conflicts with its reliability and performance [1–4].

In this regard, methods for controlling large systems, combining the simplicity of centralized control and the reliability of local control, are currently being intensively developed [5–7].

In line with this research area in the works [8–10], an evolutionary approach to the management of complex systems was proposed, which reproduces the property of

V. Taratukhin et al. (Eds.): ICID 2022, CCIS 1767, pp. 27–39, 2023.
https://doi.org/10.1007/978-3-031-32092-7_3

large ecosystems to maintain the ability for sustainable development in a wide range of changes in the parameters of the external environment and internal elements. This approach is formulated as a concept of multi-alternative structure and functioning of a complex system, and is based on a number of principles: the principle of multi-level control system, the principle of diversity and division of functions at each level of the system, and the principle of modularity [11].

In the proposed work, an example of applying these principles to control a complex dynamic system with a reference model is considered. The presence of such a model in an explicit or implicit form predetermines the desired state of the object, which is, ultimately, the goal of control [12–20].

Known control schemes with a reference model are characterized by cumbersome implementation – in the case of using parametric adaptation [21–24], or a small depth of regulation – in the case of signal adaptation [25–27]. The article will show that the construction of a control system based on the principles of multiple alternatives allows, under certain conditions, eliminating the indicated shortcomings of known schemes and ensuring the high quality of the system, combined with the simplicity of control algorithms.

The claimed advantage of the proposed approach arises from the fact that the original object of a high order is structurally and physically represented in the form of elements of a low dynamic order. A simple signal adaptation subsystem is built for each selected element, and then the resulting subsystems are combined into a hierarchical structure containing two parallel channels: a channel from the object's elements and a channel from the corresponding cascades of the reference model. The resulting structure, which combines the block structure and multi-level control, eliminates the accumulation of adaptation errors in each cascade of the system, and removes the contradiction between the requirements for the stability of the system as a whole and its complexity [28].

The article has the following structure:

- formal statement of the problem;
- analysis of adaptive properties of known and proposed control methods;
- obtaining the conditions for stabilizing the system when changing the object's parameters;
- presentation of the method for solving the problem set on the basis of the principles of multiple alternatives;
- numerical modeling and comparison of known and proposed control schemes on a specific numerical example.

2 Problem Statement

Let's consider the object:

$$\dot{x}(t) = Bx(t) + Nu(t);$$
$$\mathrm{y(t) = A\ x(t)}, \qquad (1)$$

where $x(t) = [x(t)_1 \ldots x(t)_n]^{\mathrm{T}}$ is the object state vector; $y(t)$ – controlled (output) value; $u(t)$ – control action (argument t is omitted below); B – characteristic matrix of the object, $[n \times n]$; N – control matrix, $[n \times 1]$; A – output matrix, $[1 \times n]$. Matrices B, N and A of the object contain non-stationary elements that change in time.

Let's apply to this object a single-level signal adaptation scheme with an explicit reference model, Fig. 1 [16]:

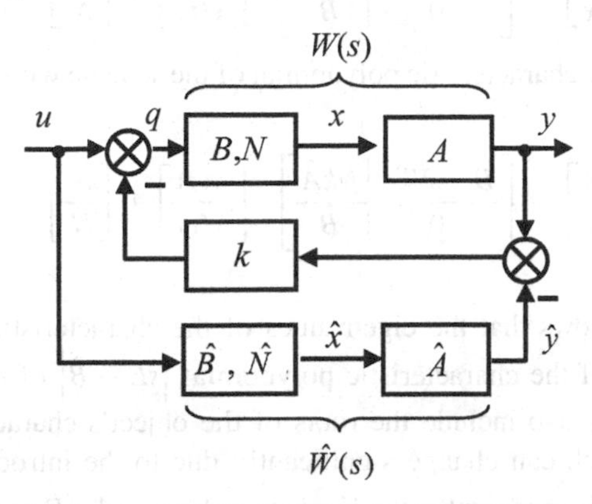

Fig. 1. Block diagram of a system with single-level signal adaptation and an explicit reference model.

The reference model of the object is indicated in Fig. 1 by matrices $\hat{B}, \hat{N}, \hat{A}$ and quantities $\hat{x}, \hat{y}$ having names similar to the object eq. (1). The coefficient k is a scalar constant forming a closed loop of signal adaptation; ε – adaptation error. $W(s)$ and $\hat{W}(s)$ are the transfer functions of the object and the reference model, respectively.

The task is to reproduce the reference behavior $\hat{y}$ at the output of the object with the required regulation time t_p and error ε, asymptotically tending to zero.

3 Theoretical Analysis of the Problem

The equations of motion shown in Fig. 1 diagrams look like:

$$\begin{cases} \dot{x} = Bx + N(u - k \cdot \varepsilon); \\ y = Ax, \\ \dot{\hat{x}} = \hat{B}\hat{x} + \hat{N}u; \\ \hat{y}(t) = \hat{A}\hat{x}; \\ \varepsilon = y - \hat{y}. \end{cases} \tag{2}$$

The system of Eqs. (2) indicates the fundamental possibility of signal compensation for variations in the matrices B, N and A of the object by the term $k\varepsilon$, which is included in the general control $q = u - k\varepsilon$.

Substituting from (2) the expressions for y and $\hat{y}$ into the equation for ε, and then ε into the equation for $\dot{x}$, we get:

$$\begin{cases} \dot{x} = (B - NkA)x + N(k\hat{A}\hat{x} + u); \\ \dot{\hat{x}} = \hat{B}\hat{x} + \hat{N}u, \end{cases} \tag{3}$$

Let us analyze the properties of scheme (2), (3). Let us rewrite (3) in the form:

$$\left[\frac{\dot{x}}{\dot{\hat{x}}}\right] = \left[\begin{array}{c|c} B - NkA & Nk\hat{A} \\ \hline 0 & \hat{B} \end{array}\right] \cdot \left[\begin{array}{c} x(t) \\ \hat{x}(t) \end{array}\right] + \left[\frac{N}{\hat{N}}\right] \cdot u. \tag{4}$$

As a result, we find the characteristic polynomial of the scheme with a single-level signal adaptation:

$$\left[\frac{\dot{x}}{\dot{\hat{x}}}\right] = \left[\begin{array}{c|c} B - NkA & Nk\hat{A} \\ \hline 0 & \hat{B} \end{array}\right] \cdot \left[\begin{array}{c} x(t) \\ \hat{x}(t) \end{array}\right] + \left[\frac{N}{\hat{N}}\right] \cdot u. \tag{5}$$

$E = diag(1 \ldots 1)$.

Expression (5) shows that the eigenvalues of the characteristic matrix of scheme (2) repeat the roots of the characteristic polynomial $\left|sE - \hat{B}\right|$ of the reference model without changes, and also include the roots of the object's characteristic polynomial $|sE - B + NkA|$, which can change significantly due to the introduction of the feedback coefficient k. This implies that the Hurwitz polynomial $\left|sE - \hat{B}\right|$, i.e. the reference model, is chosen to be known to be stable with some value of the maximum root of this polynomial $\hat{\lambda}_{\max} < 0$.

To analyze the influence of the coefficient k on the closed adaptation loop stability, we represent the matrices of the object (1) in the Frobenius form:

$$B = \left[\begin{array}{cc} \mathbf{0} & \mathbf{E}_{n-1} \\ \hline \multicolumn{2}{c}{-\mathrm{c}} \end{array}\right]; \quad N = [0 \ldots 0 \ 1]^{\mathrm{T}}; \quad A = [b_0 \ldots b_m \ \mathbf{0}], \tag{6}$$

where $[b_0 \ldots b_m \ \mathbf{0}]$ and $\mathbf{c} = [c_0 \ldots c_{n-1}]$– respectively, the coefficients of the polynomials of the numerator and denominator of the object's transfer function *W(s)* in Fig. 1:

$$W(s) = \frac{y(s)}{q(s)} = A(s\mathrm{E} - B)^{-1} N, \tag{7}$$

or

$$W(s) = \frac{B(s)}{C(s)} = \frac{b_m s^m + b_{m-1} s^{m-1} + \ldots + b_1 s + b_0}{s^n + c_{n-1} s^{n-1} + \ldots + c_1 s + c_0}; \quad m < n. \tag{8}$$

Rewriting the polynomial $|sE - B + NkA|$, taking into account (6), (7) and (8), we get:

$$\begin{aligned} |sE - B + NkA| &= s^n + c_{n-1} s^{n-1} + c_{n-2} s^{n-2} + \ldots \\ &+ (c_m + b_m k) s^m + (c_{m-1} + b_{m-1} k) s^{m-1} + \ldots + (c_1 + b_1 k) s + c_0 + b_0 k. \end{aligned} \tag{9}$$

Expression (9) shows that the one-level signal adaptation applied to the object (1) is accompanied by an increase in the initial coefficients c_i of its characteristic polynomial to the values $c_i + b_i k$, $(i = 0 \ldots m)$. Note that not all, but only the first $m + 1$coefficients $\mathbf{c}$ are subjected to such a change.

Such a partial and disproportionate change in the coefficients of the original characteristic polynomial $|sE - B|$ of the object, in the general case, leads to an unpredictable displacement of the matrix eigenvalues $B - NkA$ in the complex plane of their values, which conflicts with the desire to improve the quality of adaptation by increasing k.

Let's consider several specific cases.

3.1 First Order Object

Integrator. For such an object, the characteristic polynomial of the signal adaptation circuit has the form:

$$|sE - B + NkA| = s + b_0 k. \tag{10}$$

In this case, the adaptation loop for any non-negative values of k retains its stability. However, as a result of the introduction of the adaptation loop, the integrating properties of the object are lost, and it acquires the properties of an inertial link. In this regard, the analysis of this case is reduced to the analysis of objects in the form of an inertial link.

Inertial Link. Type of the characteristic polynomial of the signal adaptation contour:

$$|sE - B + NkA| = s + c_0 + b_0 k. \tag{11}$$

The loop feedback coefficient k is included only in the free term $c_0 + b_0 k$ of this polynomial, and it is always possible to choose a sufficiently large k for which the inequality $-(c_0 + b_0 k) << \hat{\lambda}_{max}$ is satisfied, i.e. the adaptation scheme will provide a system of dominant roots of its characteristic polynomial that is almost identical to the roots of the reference model polynomial $\left|sE - \hat{B}\right|$ in a wide range of changes in the object's parameters c_0 and b_0.

3.2 Second Order Object

For an object of the second order, the characteristic polynomial of the signal adaptation scheme has two implementation options – for an object with forcing:

$$|sE - B + NkA| = s^2 + (c_1 + b_1 k)s + c_0 + b_0 k, \tag{12}$$

and for an object without forcing:

$$|sE - B + NkA| = s^2 + c_1 s + c_0 + b_0 k. \tag{13}$$

Let us use the analytical expression for the roots $s_{1,2}$ of the polynomial (12):

$$s_{1,2} = -\frac{c_1 + b_1 k}{2} \pm \sqrt{\frac{(c_1 + b_1 k)^2}{4} - (c_0 + b_0 k)}. \tag{14}$$

It follows from (14) that for case (12) the choice of a sufficiently large k ensures the dominance of the roots of the reference model polynomial $\left|sE - \hat{B}\right|$ without losing the adaptation loop stability.

In the case (13), $b_1 = 0$ and expression (14) takes the form:

$$s_{1,2} = -\frac{c_1}{2} \pm \sqrt{\frac{c_1^2}{4} - (c_0 + b_0 k)}, \tag{15}$$

from which it follows that as the coefficient k increases, the roots become complex, and their imaginary part increases. The coefficient k does not affect the real part of the roots $s_{1,2}$. It follows from this that with an increase in k, the frequency of natural oscillations of the circuit increases, and can be taken out of the operating frequency range of the entire control system.

The stability of the contour is still preserved for any value of k.

Let us analyze the possibility of suppressing the resonant frequencies of the circuit in the complete adaptation scheme for the case (13).

The transfer function of the complete adaptation scheme (see Fig. 1) is:

$$W_{\mathrm{a}}(s,k) = \frac{y(s)}{u(s)} = \frac{W(s)(k\hat{W}(s)+1)}{1+kW(s)}. \tag{16}$$

Limit of expression (16) for k tending to infinity:

$$\lim_{k\to\infty} W_a(s,k) = \hat{W}(s) \tag{17}$$

indicates the fundamental – theoretical ability of the scheme in Fig. 1 to reproduce the properties of the reference model.

Let's find a quantitative – practical assessment of the values of k, at which the resonant oscillations of the adaptation scheme noted above will be suppressed to a given level, for example, (−20) dB. Let's take an object:

$$W(s) = \frac{1}{s^2+0.2s+1} \tag{18}$$

with pronounced oscillatory properties (the damping factor for this object is 0.1), and the reference model:

$$\hat{W}(s) = \frac{1}{s^2+2s+1} \tag{19}$$

with real roots of the characteristic polynomial, i.e. with the monotonous nature of the transient process at its output.

As a result of substituting (18) and (19) into (16), we obtain $W_{\mathrm{a}}(s)$, and then the frequency and transient characteristics of the adaptation scheme for values of $k = 1\ldots100$, Fig. 2.

With a value of $k = 100$, the transient response of the adaptation scheme visually coincides with the transient response $\hat{y}(t)$ of the reference model. The resonant surge in the amplitude frequency response of the scheme does not exceed (−20) dB.

Thus, for objects of the second order, the considered single-level signal adaptation scheme can be implemented with the values of the feedback coefficient k of the adaptation loop without restrictions on the stability criterion.

3.3 Third Order Object

For the third order object, the characteristic polynomial of the signal adaptation circuit for the variant without forcing has the form:

$$|sE - B + NkA| = s^3 + c_2s^2 + c_1s + c_0 + b_0k. \tag{20}$$

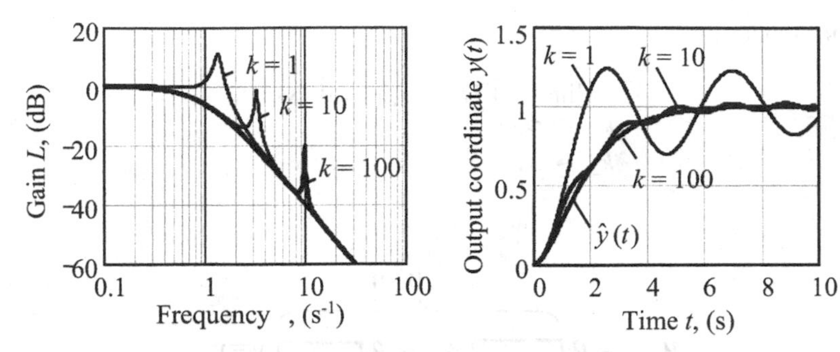

Fig. 2. Amplitude frequency responses $L(\omega)$ and transient responses $y(t)$ of the adaptation scheme for various values of the coefficient k.

The strongest condition for the realization of this variant is the stability condition. According to the Hurwitz criterion for (20), we obtain:

$$c_2 c_1 > c_0 + b_0 k. \tag{21}$$

Condition (21) for a non-stationary object is practically unsatisfiable not only because of the a priori unknown change in its parameters c_2, c_1, c_0, b_0, but also because of the fundamental contradiction with the principle of operation of the considered signal adaptation scheme: improving the adaptation quality requires increasing the coefficient k, but increasing k makes it difficult to fulfill the stability condition (21).

Similar difficulties arise for variants of the object with forcing. In some cases, the situation described can be mitigated by the fact that in real objects, the number of dominant roots of the characteristic polynomial is usually small, and such objects can be considered as dynamic links of the second and even first orders.

4 Solution Method Based on the Principles of Multiple Alternatives

The presented analysis leads to the conclusion that for the effective application of the considered signal adaptation scheme, it is necessary to physically separate the sub-objects of the first and second orders from the complete high-order object, and to form a separate adaptation scheme for each of these selected parts. In other words, it is possible to eliminate the fundamental limitations of signal adaptation identified above by using a block representation of the control object and constructing an appropriate multilevel signal adaptation scheme for each block with a dynamic order no higher than the second. The structure of such a circuit for a third-order object is shown in Fig. 3. In Fig. 3: $W_1(s)$, $W_2(s)$ – transfer functions of cascades of the first and second orders in the object, respectively; $\hat{W}_1(s)$, $\hat{W}_2(s)$ – similar transfer functions of the reference model cascades; k_1, k_2 – adaptation loop coefficients.

The transfer function $\tilde{W}(s)$ of the structure in Fig. 3:

$$\tilde{W}(s, k_1, k_2) = \frac{y(s)}{u(s)} = \frac{W(s)(k_2\hat{W}_2(s) + 1)}{(1 + k_1 W_1(s))(1 + k_2 W_2(s))} + \frac{W_1(s)k_1\hat{W}(s)}{1 + k_1 W_1(s)}. \tag{22}$$

It is obvious that:

$$\lim_{\substack{k_1 \to \infty \\ k_2 \to \infty}} \tilde{W}(s, k_1, k_2) = \hat{W}(s). \tag{23}$$

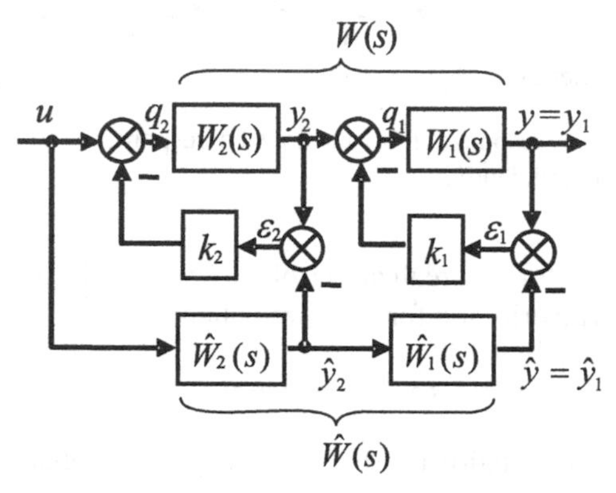

Fig. 3. Block diagram of a system with two-level signal adaptation and an explicit reference model.

It is essential that multilevel signal adaptation is not a trivial serial connection of single-level cascades. In this case, an adaptation error would accumulate in each cascade.

As follows from Fig. 3, the relationship between the adaptation levels is divided into two channels: the reference channel in each stage generates reference controls $\hat{y}_2$ and then $\hat{y}_1$, regardless of the true state of the object, i.e. no error. Therefore, the error ε_2 contained in the signal y_2 will enter the signal q_1, then ε_1 and will be compensated together with the first level own error.

The motion equations of lower levels of adaptation cannot be described using a single transfer function of the form (16), but have the form, for example, for Fig. 3:

$$y(s, k_1) = \frac{W_1(s)}{1 + k_1 W_1(s)} y_2(s) + \frac{W_1(s) k_1 \hat{W}_1(s)}{1 + k_1 W_1(s)} \hat{y}_2(s). \tag{24}$$

With an increase in k_1, we come to the following result, which is equivalent to (17):

$$\lim_{k \to \infty} y(s, k_1) = \hat{W}_1(s) \hat{y}_2(s). \tag{25}$$

The two-level signal adaptation scheme shown in Fig. 3, as already noted, cannot contain a series-connected integrator due to the closure of each stage with feedback. Therefore, the scheme in Fig. 3 does not have the properties of astatism, i.e. it always has a static error between y and u. To eliminate this error, the signal adaptation scheme should be covered by a common feedback, and a proportional-integral (PI) or other controller with an integrating link should be included in the direct channel of the system (see the example in paragraph 5).

5 Numerical Simulation Results

For a numerical example, we will use a third-order object, which can be divided into two cascades – first and second orders, and in the general case, the second-order cascade has complex roots of the characteristic polynomial:

$$W(s) = \frac{b_0^1}{s + c_0^1} \cdot \frac{b_0^2}{s^2 + c_1^2 s + c_0^2}. \tag{26}$$

The values of non-stationary parameters of the object are presented in the Table 1.

Table 1. Options for non-stationary object parameters

Variant	c_0	c_1	c_2	b_0	c_0^1	b_0^1	c_0^2	c_1^2	b_0^2
v1	0.04	0.22	0.3	0.1	0.2	1	0.2	0.1	0.1
v2	2	2.2	10.2	1	10	1	0.2	0.2	1
v3	5	2	5.2	1	5	1	1	0.2	1

As a reference model, we choose two links of the first and second orders with monotonic transient responses and a total control time $t_r = 6.4$ s:

$$\hat{W}(s) = \frac{1}{s + 1} \cdot \frac{1}{s^2 + 2s + 1}. \tag{27}$$

The transient characteristics of the variants of the object and the reference model are shown in Fig. 4 and demonstrate significant changes in the object properties with the changes in the parameters indicated in the table.

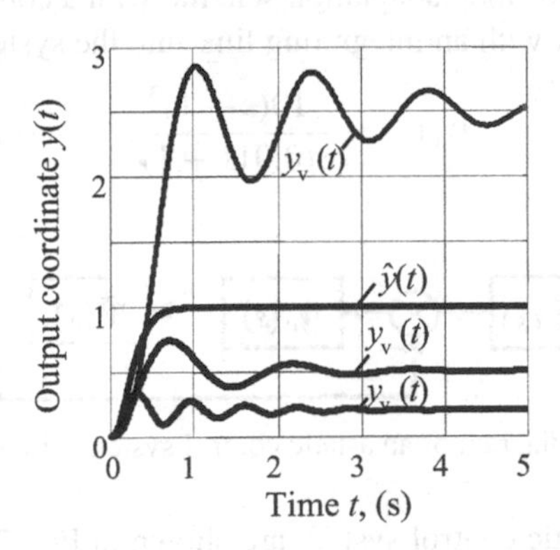

Fig. 4. Transient characteristics of variants of the object and the reference model.

We implement a two-level signal adaptation scheme according to Fig. 3, in which, based on the above analysis, we take $k_1 = k_2 = 100$. Transient processes are shown in

Fig. 5. For a visual comparison, Fig. 5 shows the result of a one-level adaptation of the v1 variant according to the scheme in Fig. 1. In this scheme, the stability condition (21) limits the coefficient k of the variant v1 to a critical value:

$$k < \frac{c_2 c_1 - c_0}{b_0} = \frac{0.3 \cdot 0.22 - 0.04}{0.1} = 0.26. \tag{28}$$

To ensure the margin of stability in the single-level adaptation scheme, the value of $k =$ 0.15 is chosen.

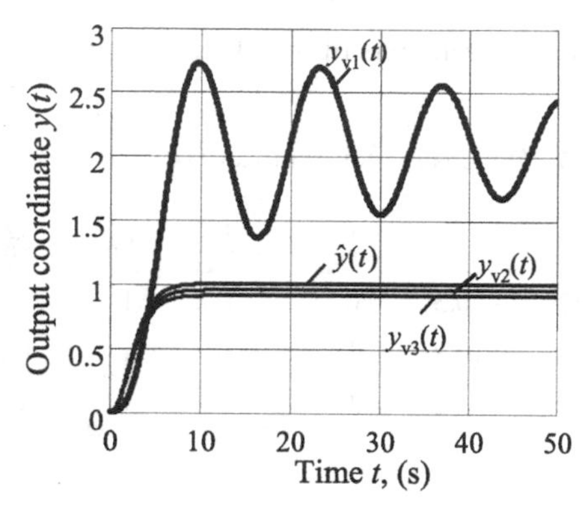

Fig. 5. Transient responses for the reference model ($\hat{y}(t)$) and object variants in single-level ($y_{v1}(t)$) and two-level ($y_{v2}(t)$, $y_{v3}(t)$) signal adaptation schemes.

Figure 5 confirms the practical unsuitability of one-level adaptation for this example (this was to be expected due to the extremely small coefficient $k = 0.15$), and also illustrates the high efficiency of the two-level adaptation scheme.

At the same time, Fig. 5 shows the theoretically predicted existence of a static error in the values $y_{v2}(t)$ and $y_{v3}(t)$ relative to the reference process $\hat{y}(t)$. In order to eliminate this error, we cover the two-loop adaptation scheme with a common negative feedback and introduce a controller with an integrating link into the system, Fig. 6:

$$W_c(s) = \frac{10(s+1)^3}{s(0.01s+1)^2}. \tag{29}$$

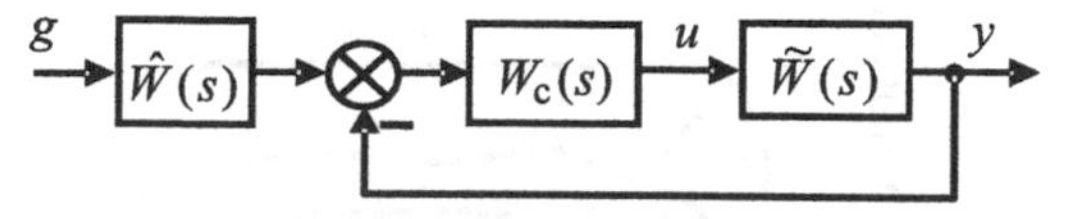

Fig. 6. Structural diagram of an astatic control system with signal adaptation.

Transients in the astatic control system are shown in Fig. 7 and differ by fractions of a percent from the reference process for all variants of non-stationary parameters indicated in the table.

Note that the regulation time in the complete control system corresponds to the set value $t_r = 6.4$ s (see Fig. 7).

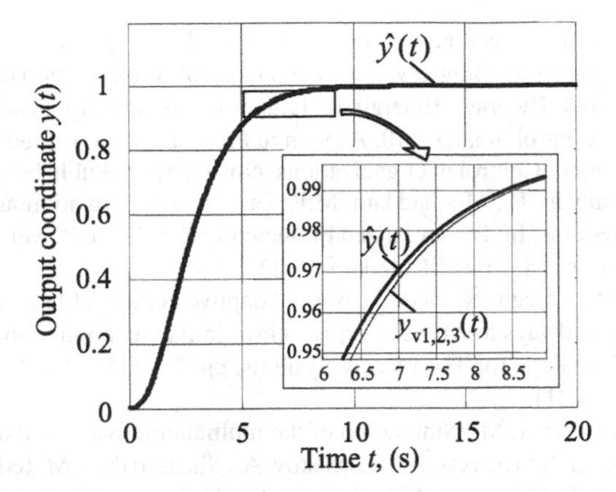

Fig. 7. Transients in a complete control system with multilevel signal adaptation and an additional regulator.

6 Conclusion

The results of the study of a complex signal adaptation system with an explicit reference model presented in the paper allow us to draw the following conclusions:

- the traditionally considered narrow operating range of signal adaptation can be significantly expanded if the non-stationary control object is structurally and physically represented in the form of series-connected cascades with the dominant dynamic order of each of them not higher than the second;
- the specified representation of the object makes it possible to implement a multilevel signal adaptation scheme, which is a hierarchically subordinate connection of single-level schemes with two parallel channels: a channel with object stages and a channel with the corresponding stages of the reference model. Such an organization of adaptation levels excludes the accumulation of adaptation errors of each cascade;
- the division of the adaptation scheme into parts of a low order fundamentally eliminates the contradiction between the quality of the adaptation scheme and its stability;
- the simulation results confirm the high efficiency of the proposed signal adaptation scheme and thus demonstrate the constructive nature of the evolutionary concept of multiple alternatives used in this work.

References

1. Kolesnikov, A.A.: Introduction of synergetic control. In: Proceedings of the American Control Conference ACC-2014, pp. 3013–3016. IEEE, Curran Associates, Portland, Oregon, USA (2014)
2. de Souza, G.F.M., Melani, A.H.D.A., Michalski, M.A.D.C., Da Silva, R.F. (eds.): Reliability Analysis and Asset Management of Engineering Systems. Elsevier, Straive, India (2021)
3. Valeev, S., Kondratyeva, N.: Process Safety and Big Data. Elsevier, SPi Global, India (2021)

4. Najafi, M.: Stabilization techniques for large input-delay systems with uncertainty. In: Khooban, M.-H., Tomislav Dragicevic, T. (eds.) Control Strategy for Time-Delay Systems, Part I: Concepts and Theories. Emerging Methodologies and Applications in Modelling, Identification and Control, pp. 83–110. Academic Press, London, United Kingdom (2021)
5. Novikov, D.: Theory of Control in Organizations. Nova Science Publishers, New York (2013)
6. Sioshansi, F.P., Zamani, R., Moghaddam, M.P.: Energy transformation and decentralization in future power systems. In: Decentralized Frameworks for Future Power Systems, pp. 1–18. Academic Press, London, United Kingdom (2022)
7. Hashemipour, S.H., Vasegh, N., Sedigh, A.K.: Adaptive control of large-scale systems with long input and state delays and time-varying delays in the uncertain nonlinear interconnections. In: Control Strategy for Time-Delay Systems, pp. 277–313. Academic Press, London, United Kingdom (2021)
8. Podvalny, S.L., Vasiljev, E.M.: Simulation of the multialternativity attribute in the processes of adaptive evolution. In: Kravets, A., Bolshakov, A., Shcherbakov, M. (eds.) Cyber-Physical Systems: Advances in Design & Modelling. Studies in Systems, Decision and Control, vol. 259, pp. 167–178. Springer, Cham (2020)
9. Podvalny, S.L., Vasiljev, E.M.: A multi-alternative approach to control in open systems: origins, current state, and future prospects. Autom. Remote. Control. **76**(8), 1471–1499 (2015)
10. Podvalny, S., Vasiljev, E.: Multi-alternative control of large systems. In: MATEC Web of Conferences. "13th International Scientific-Technical Conference on Electromechanics and Robotics "Zavalishin's Readings" – 2018", vol. 161, 2003. EDP Sciences, Les Ulis, France (2018)
11. Podvalny, S.L., Vasiljev, E.M.: Evolutionary principles for construction of intellectual systems of multi-alternative control. Autom. Remote. Control. **76**(2), 311–317 (2015)
12. Annaswamy, A.M., Fradkov, A.L.: A historical perspective of adaptive control and learning. Annu. Rev. Control. **52**, 18–41 (2021)
13. Lui, D.G., Petrillo, A., Santini, S.: Distributed model reference adaptive containment control of heterogeneous multi-agent systems with unknown uncertainties and directed topologies. J. Franklin Inst. **358**(1), 737–756 (2021)
14. Xie, J., Yang, D., Zhao, J.: Composite anti-disturbance model reference adaptive control for switched systems. Inf. Sci. **485**, 71–86 (2019)
15. Aguila-Camacho, N., Gallegos, J.A.: Switched fractional order model reference adaptive control for unknown linear time invariant systems. IFAC-PapersOnLine **53**(2), 3731–3736 (2020)
16. Ristevski, S., Dogan, K.M., Yucelen, T., Muse, J.A.: Transient performance improvement in reduced-order model reference adaptive control systems. IFAC-PapersOnLine **52**(29), 49–54 (2019)
17. Hou, Q., Dong, J.: Adaptive fuzzy reliable control for switched uncertain nonlinear systems based on closed-loop reference model. Fuzzy Sets Syst. **385**, 39–59 (2020)
18. György, K., Dávid, L.: Comparison between model reference discrete time indirect and direct adaptive controls. Procedia Manuf. **22**, 444–454 (2018)
19. Shiota, T., Ohmori, H.: Variable reference model for model reference adaptive control system. IFAC-PapersOnLine **48**(14), 72–75 (2015)
20. Dogan, K.M., Yucelen, T., Muse, J.A.: Adaptive control systems with unstructured uncertainty and unmodelled dynamics: a relaxed stability condition. Int. J. Control **95**(8), 2211–2224 (2022)
21. Yang, Z., Jin, Z.: Modeling and specifying parametric adaptation mechanism for self-adaptive systems. In: Zowghi, D., Jin, Z. (eds.) Requirements Engineering. CCIS, vol. 432, pp. 105–119. Springer, Heidelberg (2014). https://doi.org/10.1007/978-3-662-43610-3_9
22. García-Valls, M., Perez-Palacin, D., Mirandola, R.: Pragmatic cyber physical systems design based on parametric models. J. Syst. Softw. **144**, 559–572 (2018)

23. Gerasimov, D.N., Pashenko, A.V.: Robust adaptive backstepping control with improved parametric convergence. IFAC-PapersOnLine **52**(29), 146–151 (2019)
24. Gerasimov, D.N., Nikiforov, V.O.: Augmented error based adaptive control with improved parametric convergence. IFAC-PapersOnLine **55**(12), 67–78 (2022)
25. Flores-Pérez, A., Grave, I., Tang, Y.: Comparative analysis of passive algorithms in adaptive control. Int. J. Adapt. Control Signal Process. **28**(10), 1043–1053 (2014)
26. Dadenkov, D.A., Kazantsev, V.P.: On the synthesis of passive–adaptive systems for electric drive control. Russian Electr. Eng. **86**(11), 661–666 (2015). https://doi.org/10.3103/S1068371215110024
27. Khasanov, O., Khasanov, Z., Khasanova, N.: The reference model control algorithms with signal adaptation for multiply connected objects of a robotic technological complex. In: International Conference on Electrotechnical Complexes and Systems (ICOECS), pp.1–5. IEEE, USA (2020)
28. Podvalny, S.L., Vasiljev, E.M.: Multi-alternative stabilization of structurally unstable objects. In: "2015 International Conference "Stability and Control Processes" in Memory of V.I. Zubov (SCP)", pp. 120–122. IEEE, USA (2015)

Analysis of the Quality of Estimates in the Problem of Parametric Identification of Distributed Dynamic Processes in the Case of an Explicit and Implicit Difference Scheme

M. G. Matveev, E. A. Sirota(✉), and E. A. Kopytina

Voronezh State University, 1 Universitetskaya pl., Voronezh 394018, Russia
atoris@list.ru

Abstract. The article considers the solution of the problem of parametric identification of atmospheric temperature in the case of the usual least squares method and the author's method of the modified least squares method. If the processes are adequately described by linear differential equations, it is convenient to switch to difference equations. In this paper, the results of modeling in the case of explicit and implicit difference schemes are discussed, the quality of estimates of parameters of the identification problem is evaluated. The results of the numerical experiment show that when the modified least square method (LSM) is applied to the parametric identification problem based on the Crank-Nicholson scheme, the best quality of model parameter estimates is obtained. The results are slightly worse for the Crank-Nicholson scheme and the implementation of the usual least squares method. The worst result as estimates was shown by the usual for the multiple autoregression model. This can be explained as follows. First, we conducted an additional study in which we showed that in the case of an explicit scheme and negative parameters of the identification problem, the convexity condition is violated, although the conservativeness condition itself is fulfilled. The reason for this may be a violation of the state of the Clock. In the case of using an implicit difference scheme, for example, Krank-Nicholson, we have shown that even in the case of negative parameters, the stability state is preserved. Secondly, the Krank-Nicholson scheme is a five-point difference scheme, which also improves the quality of estimates of the parameters of the atmospheric temperature field model.

Keywords: autoregressive model · finite difference equations · identification · LSM estimates · biased estimates of model parameters · the reduction of the dimensionality · explicit difference scheme · implicit difference scheme · conservativeness condition

1 Introduction and Problem Statement

Model studies of meteorological processes are a necessary component in forecasting systems and accounting for weather events in various practical fields: agricultural production, aviation, etc. Modeling the behavior of temperature fields in the atmosphere is an

V. Taratukhin et al. (Eds.): ICID 2022, CCIS 1767, pp. 40–50, 2023.
https://doi.org/10.1007/978-3-031-32092-7_4

important component of these studies. At the same time, models represented by parabolic differential equations are widely used, for example, convective diffusion models. Diffusion and advection are the main factors determining the dynamics of temperature fields in the atmosphere on given isobaric surfaces. With a given model structure, there is a need to evaluate its parameters, i.e. there is a problem of parametric identification. Despite a significant number of studies conducted in the field of parametric identification, including works [1–5], research in this area is relevant to this day. At the same time, almost all known works are based on statistical research methods that allow obtaining statistical estimates of the true values of the parameters of models of real dynamic objects or phenomena. However, the quality of the estimates obtained does not always meet the requirements of applications, in particular, in meteorological problems working with a particular mathematical model, especially in conditions of strong noise of the observed variables. In this regard, the problem of improving the quality of estimates of model parameters for a given model structure remains urgent. The use of a statistical approach to solving the parametric identification problem necessitates the transition from a continuous differential equation to its discrete analogues. Discrete analogs can be obtained by approximating the original differential equation with finite-difference schemes of various types defining a regular grid in a variable space, at the nodes of which multidimensional time series are formed. Time series form statistical samples of variable values necessary for the implementation of statistical identification procedures. Adequate models of multidimensional time series are the equations of multidimensional autoregression, the study of which is still fragmentary. Obviously, different schemes of finite-difference approximation will determine different structures of time series models and, as a consequence, different identification quality.

The article considers the solution of the problem of parametric identification of distributed dynamic processes. The identification parameters themselves are not constant values, but depend on time. This greatly complicates the solution of the problem. In order to obtain estimates of parametres, data classification was carried out, and then the problem of parametric identification within each class was solved.

We have considered such difference equations for which the conservativeness condition was met. The solution of the parametric identification problem was carried out on the basis of the usual least squares method, as well as the method proposed by the authors, called modified least square method LSM with the aim of improving the quality of estimates in the context of the MAPE parameter. Note that at the first stage, the case of an explicit difference scheme was considered. To identify the parameters within each class, two situations were considered: the parameters are a convex linear combination and the sum of the parameters is equal to one, but there are also negative parameters. In the second case, the quality of assessments drops markedly. It was decided to use an implicit unconditionally stable Krank-Nicholson difference scheme. And then we have to solve the problem of parametric identification based on LSM and modified LSMs.

There are many different approaches to solving this problem. One of the convenient ways to solve the problem of parametric identification is to use the time series apparatus. In [6] we showed that if the processes are adequately described by linear differential equations, then it is convenient to switch to difference equations. Also in our works, we discussed the conservativeness condition [7, 8], which consists in the equality of the sum of the parameters of the difference equation to one. In addition, the parameters of the difference equation form a convex linear combination. In [9, 10] we compared several algorithms for solving the parametric identification problem in order to improve the quality of estimates. All these algorithms are implemented based on the least squares method.

2 Research Models and Methods

In this section, we will show the models and methods of parametric identification that we are studying.

The simplest model describing the behavior of the temperature field on a given isobaric surface is the two-dimensional diffusion-advection equation:

$$\frac{\partial x}{\partial t} = D_1 \frac{\partial^2 x}{\partial l_1^2} + D_2 \frac{\partial^2 x}{\partial l_2^2} - v_1 \frac{\partial x}{\partial l_1} - v_2 \frac{\partial x}{\partial l_2}, \tag{1}$$

where D_1, D_2 are diffusion coefficients, v_1, v_2 are components of the advection velocity vector, l_1 is latitude, l_2 is longitude.

The transition to the difference form allows one to obtain the autoregressive equations.

$$y_i^{t+1} = a_1 y_{i-1}^t + a_2 y_i^t + a_3 y_{i+1}^t + \varepsilon_t, \tag{2}$$

where $\varepsilon_1, \varepsilon_2, \ldots, \varepsilon_t$ are independent random variables having the same normal distribution with zero mean and variance σ^2.

2.1 Methods for Obtaining LSM Estimates

LSM is the main method of regression analysis, which is used to estimate the unknown parameters of linear autoregression models from sample data.

In general, the LSM takes the following form:

$$\sum_t e_t^2 = \sum_t (y_i^{t+1} - (a_1 y_{i-1}^t + a_2 y_i^t + a_3 y_{i+1}^t))^2 \rightarrow \min_{a_1, a_2, a_3}. \tag{3}$$

Let y a column vector of observations of the explained variable, and Y this is $(n \times 3)$ a matrix of observations of factors (rows of the matrix - vectors of factor values in a given observation, columns - vector of values of a given factor in all observations). The matrix representation of the linear model has the form:

$$y = Ya + \varepsilon.$$

Then the vector of estimates of the explained variable and the vector of autoregression residuals will be equal to

$$\hat{y} = Ya, e = y - \hat{y} = y - Ya.$$

Differentiating this function with respect to the parameter vector a and equating the derivatives to zero, we obtain a system of equations (in matrix form) of the form:

$$(Y^T Y)a = Y^T y.$$

We now describe the measurement error model. Let the true relationship between the actual observed value and the explanatory variables take the following form:

$$\tilde{y} = \tilde{Y}\alpha,$$

where $\tilde{y}$ is the $(n \times 1)$ vector of observations of the explained variable, $\tilde{Y}$ is the $(n \times 3)$ matrix of true values of the explanatory variable, α is the (3×1) vector of regression coefficients. Values $\tilde{y}$ and $\tilde{Y}$ are not observed due to the presence of a measurement error. Instead, the values of $\tilde{y}$ and $\tilde{Y}$ are observed with additive measurement errors:

$$y = \tilde{y} + u, \quad Y = \tilde{Y} + V$$

where y is the $(n \times 1)$ vector of observed values of the explained variables which are observed with the $(n \times 1)$ measurement error vector u. Similarly, Y is $(n \times 3)$ a matrix of observed values of explanatory variables which are observed with the $(n \times 3)$ matrix V of measurement errors in Y. In this case, we can assume that the usual perturbation term is included in u without loss of generality.

The above formula can also be represented as:

$$y = \tilde{Y}\alpha + u, \quad Y = \tilde{Y} + V$$

it can be assumed that only Y is measured with measurement errors V, and u can be considered as an ordinary perturbation term in the model.

If some of the explanatory variables are measured without any measurement error, then the corresponding values of V will be zero.

Let's pretend that:

$$E(u) = 0, \quad E(uu^T) = \sigma^2 I, \quad E(V) = 0, \quad E(V^T V) = \Omega, \quad E(V^T u) = 0.$$

The following set of equations describes the measurement error model:

$$y = \tilde{Y}\alpha, \quad y = \tilde{y} + u, \quad Y = \tilde{Y} + V,$$

which can be expressed as:

$$y = \tilde{y} + u = \tilde{Y}\alpha + u = (Y - V)\alpha + u = Y\alpha + (u - V\alpha) = Y\alpha + \omega,$$

where $\omega = u - V\alpha$ is called the composite perturbation. This model resembles a conventional linear autoregression model. The basic assumption in a linear autoregression model is that the explanatory variables and perturbations are not correlated. Let's check this assumption in the model as follows:

$$E[\{Y - E(Y)\}^T \{\omega - E(\omega)\}] = E[V^T (u - V\alpha)]$$
$$= E[V^T u] - E[V^T V]\alpha = 0 - \Omega\alpha = -\Omega\alpha \neq 0.$$

Thus, Y and ω are correlated. Thus, LSM will not give an effective result.

Then the LSM is defined as:

$$a = (Y^T Y)^{-1} Y^T ya - \alpha = (Y^T Y)^{-1} Y^T (Y\alpha + \omega) - \alpha$$
$$= (Y^T Y)^{-1} Y^T \omega E(a - \alpha) = E[(Y^T Y)^{-1} Y^T \omega] \neq (Y^T Y)^{-1} Y^T E(\omega) \neq 0$$

Since Y is a random matrix correlated with ω. Thus, a becomes a biased estimate of α.

The advantages of LSM include ease of implementation and a relatively small standard deviation. The main disadvantage of LSM is the presence of bias in parameter estimates. The shift can be significant in conditions of non-stationarity, stochastic dependence of regressors and random noise.

Known methods for reducing bias worsen another important characteristic of the estimate – the standard error.

2.2 Modified Least Square Method

To reduce the estimation error, we propose a modification of the least squares method based on the condition of conservativeness of the difference scheme approximating the original differential equation. The condition of conservativeness is the fulfillment of the corresponding conservation law at the discrete level. A sign of conservativeness of the difference scheme is the equality of the sum of the parameters of the reduced difference scheme to one.

The autoregression model obtained on the basis of the given difference scheme will have the form

$$y_i^{t+1} = a_1 y_{i-1}^t + a_2 y_i^t + a_3 y_{i+1}^t \tag{4}$$

with specified initial and boundary conditions:

$$y_i^0 = c, y_{i-1}^t = b_{i-1}^t, y_{i+1}^t = b_{i+1}^t, \forall t \tag{5}$$

where i - discrete values of the spatial coordinate, t – discrete time; $a_1 + a_2 + a_3 = 1$, that is, the right part of expression (4) is a convex linear combination of temperatures.

Methods of reducing the dimension of the autoregression model are proposed, taking into account the conservativeness property of the difference scheme. So in [10], taking into account the correlation of time series in neighboring grid nodes, it was proposed to change the level of the series in the i-th node to the expression

$$y_i^t = \beta_1 y_{i-1}^t + \beta_2 y_{i+1}^t + \xi^t. \tag{6}$$

Expression (6) allows us to expect to obtain parameter estimates with a lower standard error both by reducing the dimension and by reducing the correlation of time series in non-adjacent nodes. Substituting expression (6) into (4), we get

$$y_i^t = (a_1 + a_2\beta_1) y_{i-1}^t + (a_2\beta_2 + a_3) y_{i+1}^t + + \xi_i^{k+1} - \xi_{\Sigma,}^k = \theta_1 x_{i-1}^k + \theta_2 x_{i+1}^k + \Delta\xi, \tag{7}$$

where ξ_{Σ}^{k} is a convex linear combination of interference $\xi_{i-1}^{k}, \xi_{i}^{k}, \xi_{i+1}^{k}$.

The expression (8) with the obtained estimates $\hat{\beta}$ allows us to construct a system of linear equations with respect to the parameters $\hat{\beta}$:

$$a_1 + a_2\hat{\beta}_1 = \theta_1, \quad a_2\hat{\beta}_2 + a_3 = \theta_2, \quad a_1 + a_2 + a_3 = 1. \tag{8}$$

The determinant of the system (9) $\hat{\beta}_1 + \hat{\beta}_2 - 1$ is different from zero in the case when $\hat{\beta}_1 + \hat{\beta}_2 \neq 1$.

2.3 Least Squares Method with Implicit Difference Scheme

A study was conducted in which the convexity condition of the parameters was violated [9] and one of the parameters turned out to be negative. Such results were obtained for meteorological data in the task of forecasting atmospheric temperatures. It is shown that in the case of negative parameters, the Courant condition is violated. Therefore, we decided to use the ineffectively stable Krank-Nicholson difference scheme (see Fig. 1). We use the Crank-Nicholson scheme for meteorological data. The Crank-Nicholson scheme is an unconditionally stable implicit difference scheme and can be written as a whole, as shown below:

$$(2+2r)x_i^{n+1} - rx_{i-1}^{n+1} - rx_{i+1}^{n+1} = (2-2r)x_i^{n} + rx_{i-1}^{n} + rx_{i+1}^{n}, \tag{9}$$

where $r = \frac{\tau}{h^2}$.

The Crank-Nicholson scheme can be written as:

$$-(\theta_1+\theta_2)x_{i-1}^{k+1} + (1+2\theta_1)x_i^k + (\theta_2-\theta_1)x_{i+1}^{k+1} = (\theta_1+\theta_2)x_{i-1}^{k} + (1-2\theta_1)x_i^k + (\theta_1-\theta_2)x_{i+1}^{k}$$

or

$$(1+2\theta_1)x_i^k = (\theta_1+\theta_2)x_{i-1}^{k} + (1-2\theta_1)x_i^k + (\theta_1-\theta_2)x_{i+1}^{k} +$$
$$(\theta_1+\theta_2)x_{i-1}^{k+1} + (\theta_1-\theta_2)x_{i+1}^{k+1}.$$

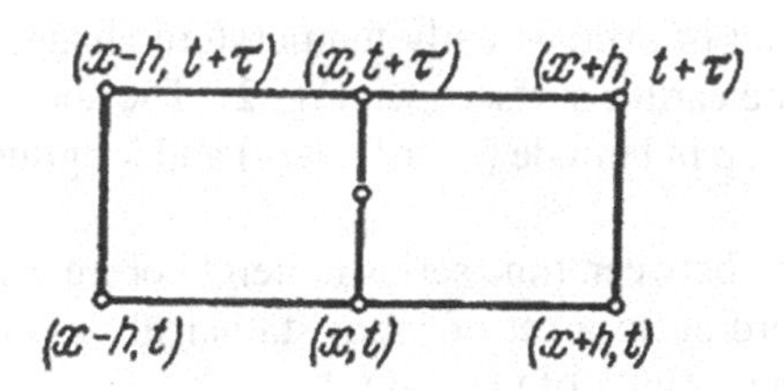

Fig. 1. Type of Crank-Nicolson difference scheme.

Divide both parts of the last inequality by $(1+2\theta_1)$, we get

$$x_i^k = \frac{(\theta_1+\theta_2)}{(1+2\theta_1)}x_{i-1}^{k} + \frac{(1-2\theta_1)}{(1+2\theta_1)}x_i^k + \frac{(\theta_1-\theta_2)}{(1+2\theta_1)}x_{i+1}^{k} + \frac{(\theta_1+\theta_2)}{(1+2\theta_1)}x_{i-1}^{k+1} + \frac{(\theta_1-\theta_2)}{(1+2\theta_1)}x_{i+1}^{k+1} \tag{10}$$

The sum of the coefficients in the right part is equal to one (this is easy to verify), that is, the condition of conservativeness is fulfilled. At the same time, if the coefficients on the right side are negative, then this will not contradict the Courant condition, unlike the previous difference scheme in our studies. Let 's introduce the notation:

$$a_1 = \frac{(\theta_1 + \theta_2)}{(1 + 2\theta_1)}; a_2 = \frac{(1 - 2\theta_1)}{(1 + 2\theta_1)}; a_3 = \frac{(\theta_1 - \theta_2)}{(1 + 2\theta_1)}.$$

Let, for example, $a_3 < 0$, the rest are positive

$$\frac{(\theta_1 - \theta_2)}{(1 + 2\theta_1)} < 0.$$

As $\theta_1 = \frac{D\Delta t}{2(\Delta z)^2}$, $\theta_2 = \frac{\vartheta \Delta t}{4\Delta z}$, then $1 + 2\theta_1 > 0$. Thus
$\theta_1 - \theta_2 < 0 => \theta_1 < \theta_2$.

It doesn't contradict anything. That is, in the case of using the Krank-Nicholson scheme, the requirement of linear convexity of the coefficients is not necessary. In the case of solving the problem of parametric identification by the method of least squares for the problem of forecasting meteorological data [6], it is proposed to conduct a preliminary stage, which consists in the formation of classes of homogeneous statistical data. Within each class, its own model will be built and the problem of parametric identification will be solved. The construction of classes of homogeneous statistics can be carried out on the basis of statistical clustering methods or on the basis of postulating physical conditions of uniformity and checking these conditions according to a measurable criterion [6]. The second direction seems more appropriate in our conditions.

To solve the problem of parametric identification in the case of the Krank-Nicholson scheme, the least squares method was used. In Eq. (10), the role of the free term for the least squares method is played by x_i^k in the left part. Unknown parameters are found for the right side of the equation with the corresponding variables.

3 Experimental Study of the Quality of Assessments

We will consider a model of average daily temperature changes in the nodes of a flat coordinate grid above the earth's surface (see Fig. 2). The source of statistical data is a web resource [12], the step in latitude (coordinate i) and longitude (coordinate j) is h = 2.5°.

There is a correlation between time series in neighboring grid nodes.

It is necessary according to a set of input statistical data representing vectors of temperature values in the vicinity of the node *ij*

$$x(t) = \left(x_{i-1,j};\ x_{i,j};\ x_{i+1,j};\ x_{i,j-1};\ x_{i,j+1}\right).$$

Obtained at various points in time, in order to distinguish classes of homogeneous statistics - for this it is necessary to determine the criterion for evaluating the uniformity of the process conditions and, accordingly, the uniformity of $M_k \in M, k = 1, .., d$.

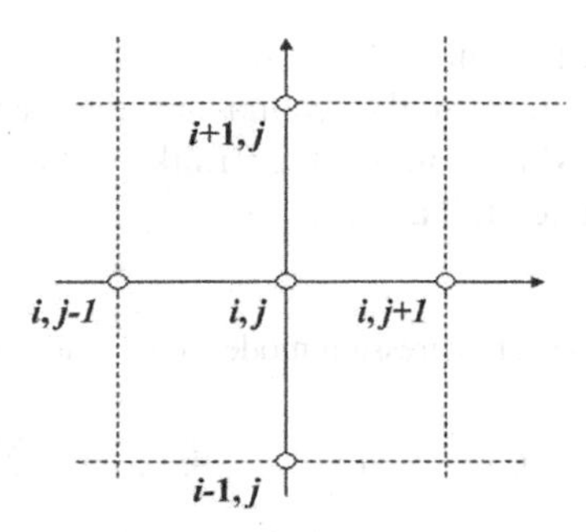

Fig. 2. View of the grid.

Atmospheric physics [8, 9] allows us to take a finite-difference estimate of the temperature gradient at each node of the spatial grid, where the components of the gradient vector are estimated by the central differences $\nabla = (\delta_i, \delta_j)\nabla = (\delta_i, \delta_j)$

$$\delta_i = x_{i+1,j} - x_{i-1,j}, \delta_j = x_{i,j+1} - x_{i,j-1}.$$

It can be assumed that the gradient calculated in this way characterizes the movement of air flows, which largely determines the current state of the atmosphere. If the gradient calculated in the vicinity of the node with coordinates *i,j*, if it does not change (or rather almost does not change) in time, then under certain assumptions it can be assumed that the state of the atmosphere in this neighborhood is stable and can be attributed to one of the classes. In other words, each class of homogeneous statistics is a combination of interval fragments of a time series with similar values of the temperature gradient. To build classes, we can propose the following procedure described in [13–15]. In each class, an autoregression model was built in each node of the temperature grid *ij*.

$$y_{ij}(t+1) = a_{ij}y_{ij}(t) + a_{i-1j}y_{i-1j}(t) + a_{i+1j}y_{i+1j}(t) + a_{ij-1}y_{ij-1}(t) + a_{ij+1}y_{ij+1}(t)$$

The parameters of the autoregression model will be determined using least square method – conventional and modified, and then we will compare the results. Classes of homogeneous statistics for meteorological data on the angle and length of the gradient vector for the usual autoregression model and the Krank-Nicholson scheme are formed. In both cases, we get 16 classes. For each class, we find unknown parameters using LSM. In the case of using the Krank-Nicholson scheme, we use a modified LSM, replacing one of the equations of the system with a value equal to one, and comparing it with the results of a conventional LSM. We compare the accuracy and adequacy of models using the coefficient of determination and the indicator of the percentage error of the MAPE forecast [16–18]

$$MAPE = \frac{1}{T}\sum\nolimits_{t=1}^{T}\left|\frac{x(t) - \hat{x}(t)}{x(t)}\right|$$

where $x(t)$ and $\hat{x}(t)$ the actual and calculated values of the levels of the MAPE series, respectively, should not exceed 10–12% [19, 20]. Below are three tables of the results of solving the parametric identification problem (each table contains the three most representative classes). The first table shows the results of solving the problem on the basis of conventional autoregression, the second - on the basis of the Crank-Nicholson scheme for a conventional LSM, the third – the Crank-Nicholson scheme and a modified least square method (see Tables 1, 2 and 3).

Table 1. Parameters of the autoregression model for the homogeneous statistics class

Class number	a_1	a_2	a_3	a_4	a_5	$\sum a$	MAPE	R^2
K6	0	0	0,66	−0,40	0,63	0,89	4,5%	0,82
K7	1,9	0	0,83	0	−1,82	0,91	5,6%	0,80
K8	−0,7	0,15	2,98	3,78	−5,40	0,81	4,8%	0,73

Table 2. Model parameters for the Crank-Nicholson scheme in the case of a conventional least square method

Class number	a_1	a_2	a_3	a_4	a_5	$\sum a$	MAPE	R^2
K6	0,5197	0,4891	−0,2707	0,5218	−0,257	1,002	3,6%	0,986
K7	0,3674	0,5662	−0,287	0,557	−0,243	0,969	3,9%	0,97
K8	0,632	0,457	−0,303	0,449	−0,291	0,944	3,76%	0,98

Table 3. Model parameters for the Krank-Nicholson scheme in the case of a modified least square method

Class number	a_1	a_2	a_3	a_4	a_5	$\sum a$	MAPE	R^2
K6	0,5135	0,465	−0,253	0,5157	−0,24	1,0012	2,9%	0,989
K7	0,345	0,559	−0,268	0,564	−0,23	0,97	3,9%	0,987
K8	0,645	0,459	−0,302	0,449	−0,299	0,952	3,2%	0,989

4 Conclusion

The results of the tables show that when a modified least square method based on the Krank-Nicholson scheme is applied to the parametric identification problem, the best quality of model parameter estimates is obtained. The results are slightly worse for the Crank-Nicholson scheme and the implementation of the usual least squares method. The worst result as estimate was shown by the usual LSM for the autoregression model.

This can be explained as follows. First, we conducted an additional study in which we showed that in the case of an explicit scheme and negative parameters of the identification problem, the convexity condition is violated, although the conservativeness condition itself is fulfilled. The reason for this may be a violation of the state of the Clock. In the case of using an implicit difference scheme, for example, Krank-Nicholson, we have shown that even in the case of negative parameters, the stability state is preserved. Secondly, the Krank-Nicholson scheme is a five-point difference scheme that also improves the quality of estimates.

References

1. Kalman, R.E.: A new approach to linear filtering and prediction problems. Transactions of the ASME **82**(1), 34–45 (1960)
2. Ben-Moshe, D.: Identification of linear regressions with errors in all variables. Econometric Theor. **37**(4), 1–31 (2020)
3. Cao, J.: Penalized nonlinear least squares estimation of time-varying parameters in ordinary differential. J. Comput. Graph. Stat. **21**, 42–56 (2012)
4. Fogler, H.R.: A pattern recognition model for forecasting. Manage. Sci. **20**, 1178–1189 (1974)
5. Liang, H.: Parameter estimation for differential equation models using a framework of measurement error in regression models. J. Am. Stat. Assoc. **103**, 1570–1583 (2008)
6. Matveev, M.G., Kopytin, A.V., Sirota, E.A., Kopytina, E.A.: Modeling of nonstationary distributed processes on the basis of multidimensional time series. Procedia Eng. **201**, 511–516 (2017)
7. Matveev, M.G., Sirota, E.A.: Analysis and investigation of the conservativeness condition in the problem of parametric identification of dynamic distributed processes. J. Phys.: Conf. Ser. **1902**, 012079 (2021)
8. Kopytin, A.V., Kopytina, E.A., Matveev, M.G.: Application of the expanded Kalman filter for identifying the parameters of a distributed dynamical system. In: Proceedings of Voronezh State University. Series: Systems Analysis and Information Technologies, vol. 3, pp. 44–50 (2018)
9. Matveev, M.G., Sirota, E.A.: Analysis of the properties of the OLS-estimators in case of elimination of multi-collinearity in the problems of parametric identification of distributed dynamic processes. Bulletin of Voronezh state University, series "System analysis and information technologies", vol. 2, pp. 15–22 (2020)
10. Matveev, M.G., Mikhailov, V.V., Sirota, E.A.: Combined prognostic model of a nonstationary multidimensional time series for constructing a spatial profile of atmospheric temperature. Inf. Technol. **22**(2), 89–94 (2016)
11. Egorshin, A.O.: Piecewise-linear identification and differential approximation on a uniform grid. In: Proceedings of the XII all-Russian conference on problems of management, 2014, p. 2807–2822. Moscow (2014)
12. NCEP/DOE AMIP II Reanalysis. URL: http://www.cdc.noaa.gov/cdc/data.ncep.reanalysis2.html
13. Govindan, R.B., Bunde, A., Havlin, S.: Volatility in atmospheric temperature variability. Phys. A: Stat. Mech. Appl. **318**(3–4), 529–536 (2003)
14. Bezruchko, B.P., Smirnov, D.A.: Modern problems of modeling from time series. In: Proceedings of the University of Sarajevo, series "Physics", vol. 6, pp. 3–27. Sarajevo (2006)
15. Guo, L.Z., Billings, S.A., Coca, D.: Identification of partial differential equation models for a class of multiscale spatio-temporal dynamical systems. Int. J. Control **83**(1), 40–48 (2010)

16. Xun, X., Cao, J., Mallick, B., Carrol, R.J., Maity, A.: Parameter estimation of partial differential equation models. J. Am. Stat. Assoc. **108**(503), 1–27 (2013)
17. Nosko, V.P.: Econometrica. Introduction to the Regression Analysis of Time Series. NFPK, Moscow (2002)
18. Gareth, J., Witten, D., Hastie, Tr., Tibshirani, R.: An Introduction to Statistical Learning. Springer (2013)
19. Samarsky, A.A.: Theory of Difference Schemes. Nauka, Moscow (1978)
20. Barseghyan, A.A., Kupriyanov, M.S., Kholod, I.I., Tess, M.D., Elizarov, S.I.: Analysis of Data and Processes. 3rd edn. BHV-Petersburg, St. Petersburg (2009)

Basics of Using Temporal Data in the Design of Project Management Information Systems

Evgeniya Kolykhalova(✉), Semyon Podvalny, and Dmitry Proskurin

Voronezh State Technical University, Voronezh, Russian Federation
evkolihalova@yandex.ru

Abstract. Today, many universities tend to use the mechanisms of project management to solve the problems of organizational and technological development. For effective change, a management mechanism is needed that ensures coordination of actions between employees of various departments. As such a mechanism, it is often proposed to use an automated information system. The main functional requirements for such a system, as a rule, are the ability to keep records of university projects, to be able to involve employees and students with the necessary competencies in projects, to track the status of projects and to minimize organizational costs and, consequently, project execution time.

Each potential project participant has a number of competencies. Each competency can be valued at some value and each competency has some value for the successful implementation of the project. Project participants, like projects, are complex objects that can be represented in the form of a multi-criteria model, in which one of the indicators must be tied to a point in time.

In this article, I would like to consider the issues of taking into account the tem-porality of data and the possibility of building a model of the state of a complex object, taking into account its possible changes in time.

Keywords: temporal data · design of information systems · project management

1 Prerequisites for the Development of a Multi-alternative System

Today, many universities tend to use the mechanisms of project management to solve the problems of organizational and technological development. Effective change requires a management mechanism that ensures coordination of actions between employees of various departments. As such a mechanism, it is often proposed to use an automated information system. The main functional requirements for such a system, as a rule, are the ability to keep records of university projects, be able to attract employees and students with the necessary competencies to participate in projects, track the status of projects and minimize organizational costs, and hence the duration of the project.

A mandatory attribute of each potential project participant is the set of competencies that he possesses. We will assume that each competency can be estimated at some value and each competency has some significance for the successful implementation of the project. Some of the competencies (attributes of the model) can be assessed

V. Taratukhin et al. (Eds.): ICID 2022, CCIS 1767, pp. 51–60, 2023.
https://doi.org/10.1007/978-3-031-32092-7_5

using information of a quantitative type, while indicators can have different units of measurement. Other indicators cannot be quantified.

Let

$$C^{+} = \{C_{i_1}^{+}, \ldots, C_{i_{s_1}}^{+}\} \quad (1)$$

be a subset of competencies at the moment of time to which requirements are imposed, and

$$C^{-} = \left\{C_{i_1}^{-}, \ldots, C_{i_{s_2}}^{-}\right\} \quad (2)$$

be the subset of competencies at the moment of time to which there are no requirements, so

$$C^{+} \cup C^{-} = C \quad (3)$$

is the set of all competencies at the moment of time that the project participant has,

$$s_1 + s_2 = s \quad (4)$$

is their number. Each of the competencies can only be assigned to one C^{+} or C^{-} class.

Obviously, there are projects for which it is important to take into account a large number of competencies of the participants, and there are projects for which the competencies of the participants do not matter. The latter, in particular, may include projects (parts of projects) involving a large amount of mechanical work, data collection, etc. Thus, the project is also a complex object and may have a number of characteristics, one of which is a set of competencies that a potential project participant must have in order to consider the possibility of participating in the project. At the same time, each necessary competence should be assessed by a certain range of values, allowing you to select project participants that best meet the needs of the project.

A set of project competencies can include both quantifiable competencies and non-quantifiable competencies. This implies the need to provide the ability to indicate the range of acceptable values of competencies, using different units of measurement. In the framework of this work, we will assume that the set of competencies and the range of acceptable values is set once for the project and does not change over time, which certainly simplifies the accounting for the set of competencies.

Let

$$P^{+} = \{P_{i_1}^{+}, \ldots, P_{i_{n_1}}^{+}\} \quad (5)$$

be a subset of competencies to which requirements are made, and

$$P^{-} = \left\{P_{i_1}^{-}, \ldots, P_{i_{n_2}}^{-}\right\} \quad (6)$$

be a subset of competencies to which requirements are not are presented, so that

$$P^{+} \cup P^{-} = P \quad (7)$$

is the set of all competencies that are important for the selection of suitable project participants,

$$n_1 + n_2 = n \tag{8}$$

is their number. Each of the competencies can only be assigned to one P^+ or P^- class (see Fig. 1).

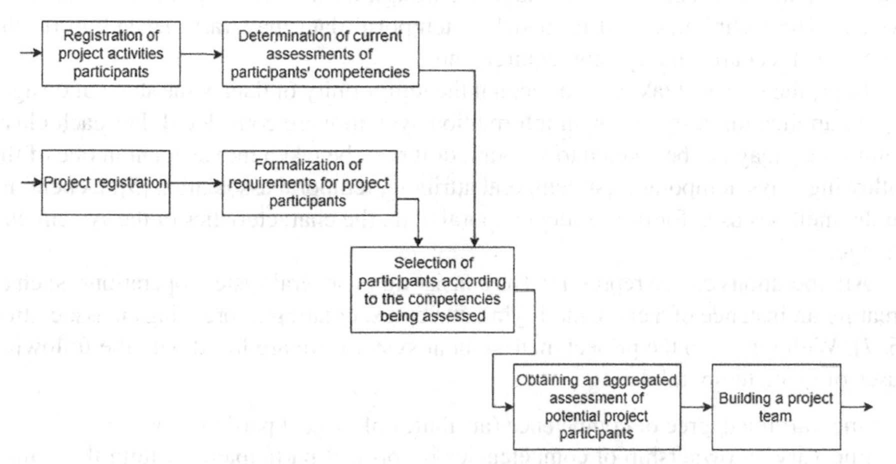

Fig. 1. The process of selecting the most suitable project participants based on the list of competencies and obtaining the final aggregated assessment of the participants.

Such a system assumes that the weights of potential project participants' competencies fall within the range of competencies acceptable for the project. However, in some cases, a situation is possible when there are no participants whose competencies fully meet the requirements of the project. In this case, it makes sense to allow the project administrator to allow the selection of participants taking into account the historical values of competencies: taking into account the dynamics of changes in the weights of required competencies over time, predicting changes in the weights of competencies in the foreseeable future within the expected life of the project, taking into account previously relevant competencies of participants, building an assumption about the presence of unaccounted for competencies acquired in the framework of previous participation in projects. This creates a multi-alternative system and allows the user to be given the opportunity to choose an alternative algorithm for selecting participants, depending on the need (willingness) to take into account the history of changes in the competencies of participants.

2 Designing Temporal Data

Under temporal data it is customary to understand any data associated with certain moments or time intervals. Often such data is also called temporal or dynamic.

Conventional relational models are primarily designed to work with static objects. However, storing the history of changing objects is in demand in many subject areas.

A class is a description of a set of objects that have the same attributes, operations, relationships, and semantics [3]. As a rule, the semantics of an instance of a class, being defined initially, does not change throughout the existence of an instance of the class, however, the attributes and relationships of class instances often tend to change over time. In cases where the projected class refers specifically to temporal data, its nature must be taken into account at all stages of designing and developing an information system. The techniques used in modeling temporal data may vary depending on the specific subject area and system requirements.

In [5] the issues of taking into account the temporality of data at the stage of designing an application program of an information system were considered. For each class, temporality may not be taken into account, or it may be taken into account in one of the following ways: temporal class, temporal attribute, temporal association [5]. Depending on the methods used for processing temporal data, the characteristics of the system will change.

All operations can be represented as consisting of several system operations, such as creating an instance of a class, modifying attributes, creating or breaking an association [6, 7]. With regard to the project management system, we are faced with the following cases of using temporal data:

- time-varying degree of competence (attribute) of project participants;
- time-varying ownership of competencies by project participants (within the framework of this project, we assume that competencies can both appear for project participants and terminate irrelevant competencies for a participant);
- time-varying weights of project participants' competencies for each specific project;
- time-varying characteristics of each specific project.

Let's consider each case in more detail.

2.1 Time-Varying Degree of Competence (Attribute) of Project Participants

One of the possible solutions to this problem is to represent attributes that tend to change over time (temporal attributes) as separate classes. In [5] they are called attribute classes (see Fig. 2).

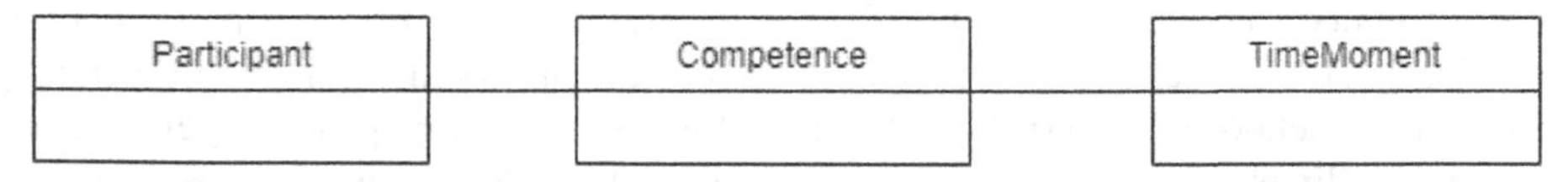

Fig. 2. Modelling of time-varying degree of competence (attribute) of project participants.

However, this scheme of working with temporal attributes is not without drawbacks. Modeling additional classes will require additional time from the developer and, accordingly, the financial costs of the customer. In addition, with sufficiently frequent changes of a large number of attributes, higher requirements will be imposed on the persistent storage devices used.

Thus, when the number of time-varying attributes is large, the above modeling method will be inefficient. In this case, it seems more attractive to associate not individual attributes with time instances, but the entire instance of the class.

2.2 Time-Varying Ownership of Competencies by Project Participants

In this case, we are talking about storing an association that changes over time. By temporal connection, we mean a connection that connects different instances of classes at different time intervals. As a solution to the problem of processing temporal associations, one can use the mechanism of creating associative objects, each of which is a separate object containing references to the identifiers of each of the participating instances [5] (see Fig. 3).

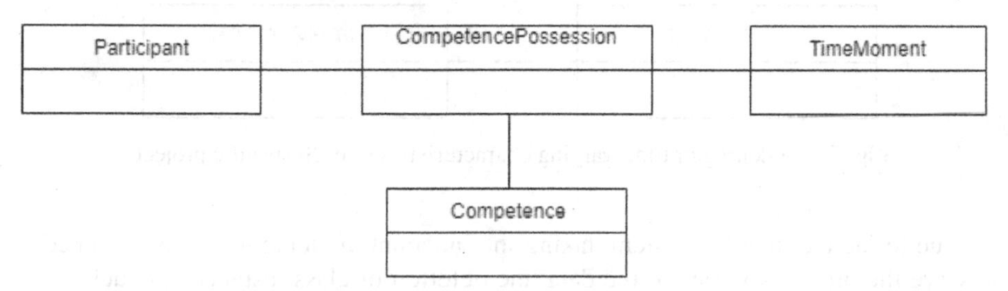

Fig. 3. Modelling of time-varying ownership of competencies by project participants.

In case of occurrence, the associative object will be the heir of the temporal class, which in turn is associated with instances of the time class.

2.3 Time-Varying Weights of Project Participants' Competencies for Each Specific Project

This case also involves the creation of attribute classes (see Fig. 4).

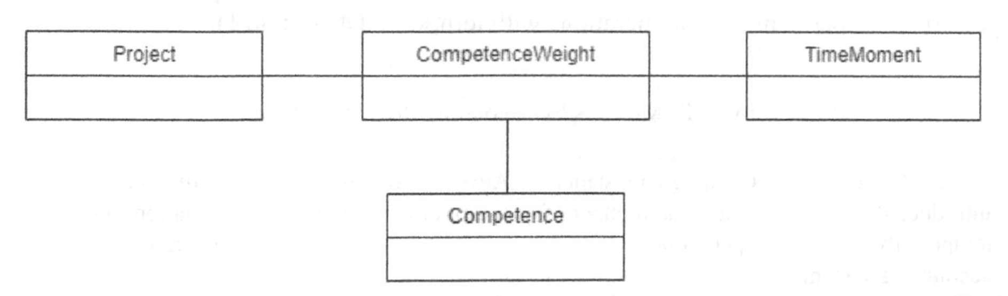

Fig. 4. Modelling of time-varying weights of project participants' competencies for each specific project.

2.4 Time-Varying Characteristics of Each Specific Project

In this case, as a rule, a large number of related characteristics of the project change simultaneously, so it is proposed to use the Snapshot approach. A snapshot of data is a representation of data at a specific point in time. The snapshot must provide time as one of its visible properties. The advantage of using the Snapshot approach is a significant reduction in the number of classes, which reduces the complexity of designing and developing an information system. In addition, this method is characterized by a significant simplification of queries to the database in comparison with the method of temporal attributes.

When using this method, the class inherits properties from some Temporal class, the main property of which is its connection with the Time class (see Fig. 5)

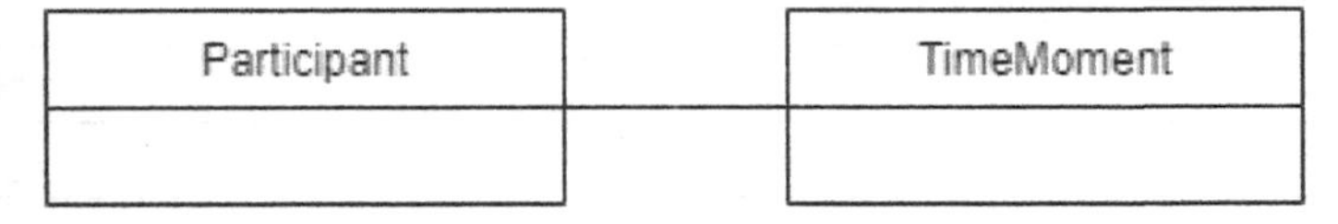

Fig. 5. Modelling of time-varying characteristics of each specific project.

Due to the fact that the system, taking into account the temporality, is designed to preserve the previous values of the data, the deletion of class instances as such is not performed in it.

The main task in modeling data temporality by the methods presented above is to design a temporal class from which other classes inherit temporal qualities.

3 Mathematical Modeling of Temporal Data

All operations can be represented as consisting of several system operations, such as creating an instance of a class, modifying attributes, creating or breaking an association [3]. Representing the operations performed on temporal data in the form of cooperation diagrams, we obtained data on the number and types of system operations that are performed to perform certain operations with temporal data (Table 1).

Table 1. Methods for accounting for temporality

The method used to introduce the temporality accounting system	Creating an instance of a class (number of operations)	Attribute modification (number of operations)	Creating an association (number of operations)
Without regard to temporality			

(continued)

Table 1. (*continued*)

The method used to introduce the temporality accounting system	Creating an instance of a class (number of operations)	Attribute modification (number of operations)	Creating an association (number of operations)
1. Create an instance of the class	1	1	–
2. Changing the class instance	–	1	–
3. Deleting an instance	1	–	–
4. Creation of an association	–	–	1
5. Rupture of association	–	–	1
Temporal class			
1. Create an instance of the class	2	2	1
2. Changing the class instance	2	3	1
3. Deleting an instance of a class	–	1	–
Temporal attribute			
1. Create an instance of the class	3	3	2
2. Changing the class instance	2	3	2
3. Deleting an instance of a class	–	2	–
Temporal association			
1. Creation of an association	2	2	3
2. Rupture of association	–	1	–

An information system without taking into account the temporality of data can be represented as a directed graph as follows:

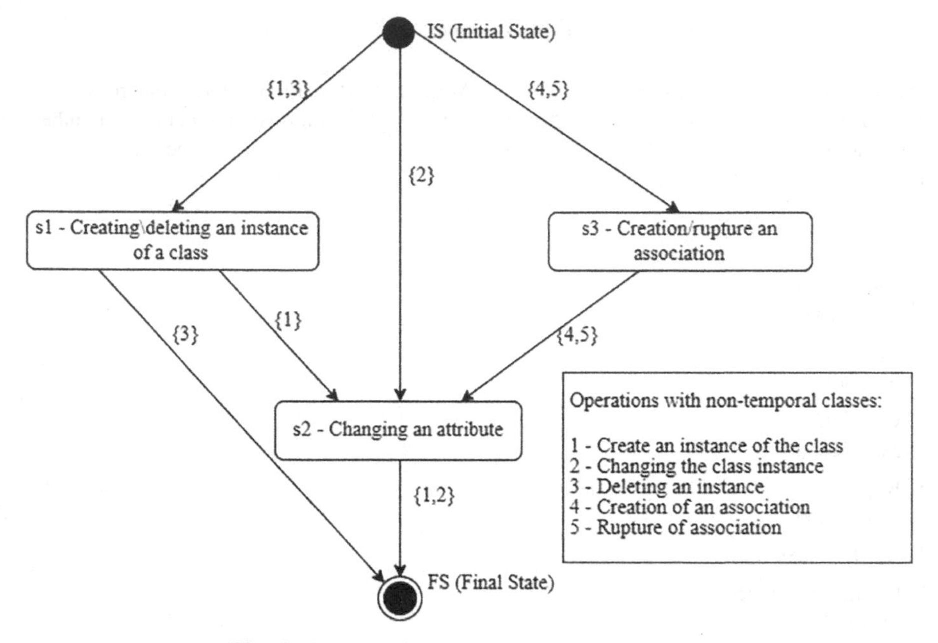

Fig. 6. System without regard to data temporality

The notion of adjacency in a directed graph presented in Fig. 6 is extended relative to a conventional directed graph [1]. In this case, we consider that the vertex s1 is adjacent to the vertex s2 when performing operation 1.

For this case, we obtain an adjacency matrix of the following form (Table 2):

Table 2. Adjacency matrix for the system without regard to temporality

	IS	S1	S2	S3	FS
IS	Ø	{1,3}	{2}	{4,5}	Ø
S1	Ø	Ø	{1}	Ø	{3}
S2	Ø	Ø	Ø	Ø	{1,2}
S3	Ø	Ø	Ø	Ø	{4,5}
FS	Ø	Ø	Ø	Ø	Ø

In the case of introducing temporality into account when designing an information system, it (the system) will correspond to the directed graph shown in Fig. 7:

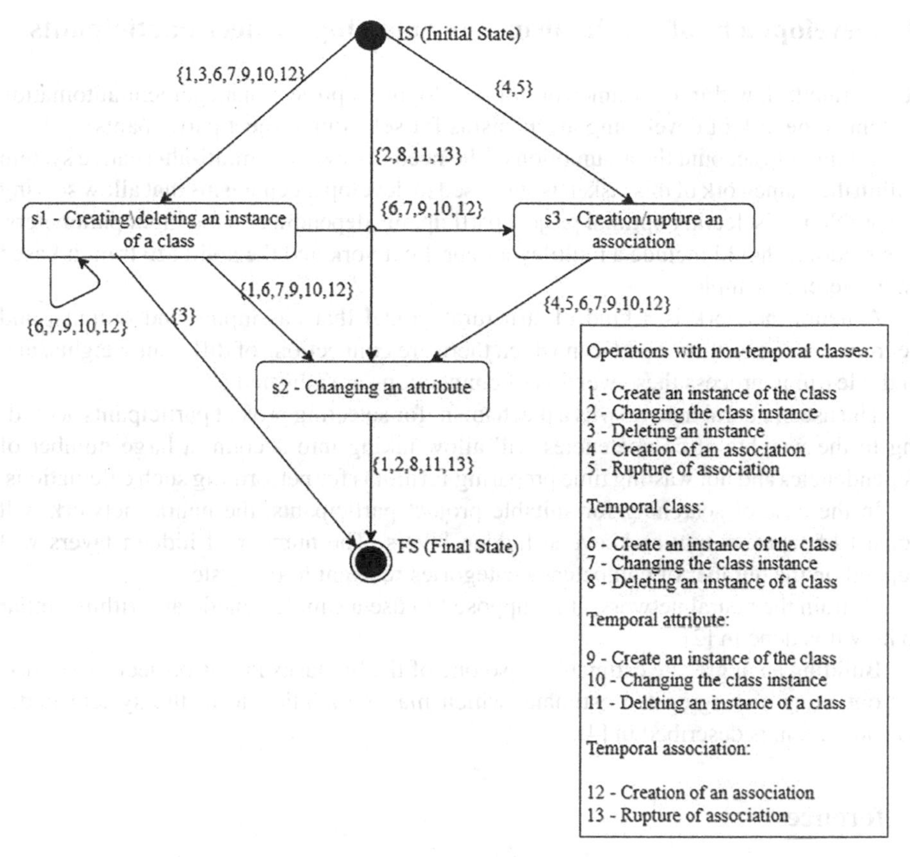

Fig. 7. Information system taking into account temporality

We get the following adjacency matrix (Table 3):

Table 3. Adjacency matrix for the system taking into account temporality

	IS	S1	S2	S3	FS
IS	Ø	{1,3,6,7,9,10,12}	{2,8,11,13}	{4,5}	Ø
S1	Ø	{6,7,9,10,12}	{1,6,7,9,10,12}	{6,7,9,10,12}	{3}
S2	Ø	Ø	Ø	Ø	{1,2,8,11,13}
S3	Ø	Ø	Ø	Ø	{4,5,6,7,9,10,12}
FS	Ø	Ø	Ø	Ø	Ø

Representing the system in the form of graphs in this way makes it possible to estimate the length of the path and the cost of resources for each operation, and, therefore, to draw conclusions about the effectiveness of the design solution when introduced into the temporality accounting system. To obtain more accurate results, it is necessary to conduct experimental studies.

4 Development of Mechanisms for Selecting Project Participants

A separate task within the framework of developing a project management automation system is the task of developing mechanisms for selecting project participants.

Taking into account the assumption of the need to develop a multi-alternative system within the framework of this task, it is supposed to develop mechanisms that allow solving the problem of selecting optimal project participants depending on the input parameters. The solution should include a multilayer neural network and the ability to train it based on a reference sample.

A neural network is a kind of structural model that has inputs and outputs, and neurons are some entities between which there are connections of different weights, and in the learning process these weights of connections are adjusted.

The use of neural networks as a mechanism for selecting project participants according to the weights of competencies will allow taking into account a large number of dependencies and not wasting time preparing formulas for performing such calculations.

In the case of searching for suitable project participants, the neural network will consist of input, output and several hidden layers. The number of hidden layers will depend on the number of competency categories relevant to the system.

To train the neural network, it is supposed to use a simple genetic algorithm similar to how it is done in [2].

Building an aggregate estimate is also one of the big tasks in this project. An option for obtaining an aggregated estimate, which may be applicable to the system under consideration, is described in [4].

References

1. Bershtein, L.S., Bozhenyuk, A.V.: The use of temporal graphs as models of complex systems. Izvestiya SFU. Technical Science. pp. 198–203
2. Emelyanov, V.V., Kureichik V.V., Kureichik, V.M.: Theory and practice of evolutionary modeling. - M.: FIZMATLIT, 432 p. (2003)
3. Larman, K.: Applying UML 2.0 and Design Patterns, 3rd ed. Per. from English. - M .: LLC "I.D. Williams" 736p. (2007)
4. Ledeneva, T.M.: Modeling of estimated systems based on the principle of multiple alternatives. Ledeneva, T.M., Podvalny, S.L.: Control and information systems technologies. - T. 57. - No. 3.1. – pp. 155–161 (2014)
5. Proskurin, D.K., Kolykhalova, E.V.: Modeling of temporal data at the level of business logic of information systems. Modern problems of informatization in modeling and social technologies: Sat. works. Issue. 15/ Ed. d.t.s., prof. O.Ya. Kravets. - Voronezh: Scientific book p. 281–283 (2010)
6. Proskurin, D.K., Kolykhalova, E.V.: Methods for designing information systems taking into account the temporality of data in the subject area, econf.rae.ru, II International Scientific Conference "Modern problems of informatization in modeling systems, programming and telecommunications", November (2009)
7. Proskurin, D.K., Kolykhalova, E.V.: Modeling of temporal information structures in the development of the information system of the university. Continuous multi-level professional education: traditions and innovations: in 3 hours. Part 2: Informatization of education. / Ed. I.S. Surovtsev; Voronezh State Architectural and Construction Univ. - Voronezh, pp. 6–8 (2011)

Digitalization as a Way to Increase the Material Utilization Rate in Mechanical Engineering

Aleksei M. Gintciak, Mikhail B. Uspenskiy, and Zhanna V. Burlutskaya(✉)

Laboratory of Digital Modeling of Industrial Systems, Peter the Great St. Petersburg Polytechnic University, St. Petersburg, Russian Federation
zhanna.burlutskaya@spbpu.com

Abstract. This work is devoted to the development of an algorithm that optimizes production planning at a discrete engineering enterprise. In the course of the work, the bottlenecks of the production process were analyzed. As part of the development of proposals to increase the resource intensity of production, existing approaches to optimizing production planning were considered. None of the existing approaches met all the functional requirements obtained as part of a series of interviews with company representatives. Based on the specifics of the company's work and technological features of production, an algorithm was developed. The developed algorithm combines production programs from several sites and distributes parts to production according to the date of need and the working time fund of individual sites. This algorithm is implemented on the basis of the method of successive approximations, which makes it possible to complete production starting from the last months of the period. Thus, the production program is distributed with compaction for the nearest period, while maintaining a reserve of production capacity for urgent orders. The developed solution is implemented in the program code and tested on various production programs of the enterprise.

Keywords: Production planning · Digitalization of production processes · Sustainable production

1 Introduction

The efficiency of the use of the enterprise's resources and its economic stability depends on compliance with the production plan (Guzman et al., 2021; Skornyakova et al., 2018). However, production planning is often carried out either manually or with partial processing of production process data, which does not allow developing realistic production programs (Kuprijanov, 2018; Athar and Janos, 2022). Digital technologies are the solution in the field of automation of production planning (Kamble et al., 2021; Alcácer and Cruz-Machado, 2019; Smirnov, 2021; Dobrinskaya, 2021).

The emergence of digital technologies has led to a global industrial transformation through the implementation of automated means of production and the transition to smart manufacturing (Chen et al., 2017). New technologies for data collection and processing have led to the creation of digital twins of products and automatization of the processes

V. Taratukhin et al. (Eds.): ICID 2022, CCIS 1767, pp. 61–70, 2023.
https://doi.org/10.1007/978-3-031-32092-7_6

of manufacturing and distribution of products (Ricciato et al., 2019). However, in the field of mechanical engineering, the use of digital technologies is mainly limited to local automation of processes (Uhlemann et al., 2017; Kritzinger et al., 2018). This feature is associated with the complexity of technological processes and a significant number of elements of the production process (Guzman et al., 2021; Qi and Tao, 2018).

In the course of this work, the practical task of using digital technologies to increase the material utilization rate is solved.

This research is carried out on the basis of a discrete engineering enterprise. The main activity of the company is the manufacture of parts of assembly units by laser and plasma cutting of metal. The distribution of parts according to the cutting cards is carried out by a specialized software solution. However, the preparation of the demand plan before sending it to the program and the processing of the results of the program is carried out manually. The purpose of this work is to develop an algorithm that optimizes the preparation of a plan for the need for details and the further distribution of the received cutting maps about the production capacities of the enterprise.

Within the framework of this work, the features of the company's production processes were analyzed based on data obtained from interviews with company representatives. Based on the information received and the analysis of existing solutions described in the articles of the Scopus database of scientific publications, functional requirements for the algorithm are formed, as well as the selection of a mathematical optimization tool is carried out. The result of the work is an algorithm for calculating the production program based on digital data on the characteristics of parts and production capacities of the enterprise.

2 Methods

As part of the study, the features of the company's production processes were analyzed based on data obtained from interviews with company representatives. Based on the interview results, the main requirements for the production program planning algorithm were identified and the existing solutions described in scientific publications of the Scopus database were analyzed. During the interview with the company's representatives, the following problems were identified:

- A large number (100+) of sheet metal types with uneven demand within different production sites. For example, an enterprise specializes in type A metal. They have large volumes of production from this type of metal. The material utilization rate for this type of metal is very good. However, this enterprise also has less commonly used metals in its production program, which cannot be effectively combined in batches.
- The material utilization rate is below the desired level due to the uneven distribution of parts on the cutting lists, as well as due to situational planning of urgent orders.
- Isolated production planning in same production sites. This means that orders placed at one enterprise are only carried out there without the possibility of merging. As a result, enterprises have to fulfill small orders, preventing them from earning on scale. Small orders statistically have lower material utilization rate.

Accordingly, based on the information received, the main tasks of project tasks were identified:

- Optimization of the material utilization rate in the long term. An increase in the material utilization rate of even a couple of percent significantly reduces production costs, which can make the enterprise cost-effective.
- Combined planning of long-term and emergency orders. Emergency orders occur as a result of loss or damage to previously manufactured parts. They must be produced as soon as possible. This overlapping of planning types lowers the material utilization rate, reducing the overall profitability of production.
- Calculation of the economic effect of the production program, taking into account logistics costs and changes in the material utilization factor. This is a completely optimization task, since increasing the material utilization rate by combining production sites leads to an increase in logistics costs.

The process of optimizing production planning should begin with the definition of the objective function: cost minimization, profit maximization, or multiple operational functions (Fahimnia et al., 2013). On the one hand, there is a clear problem of low material utilization rate, on the other hand, it is necessary to minimize logistics costs between production sites and the production of orders on time. Based on expert data from the company's representatives, it was determined that the cost of material is the most impressive item of expenditure. This is due to the increase in the market price of materials over the past few years. Thus, it is assumed that material savings should cover logistics costs, taking into account the transport accessibility of production sites.

Accordingly, one target function was identified, which is optimization of the material utilization rate in the long term. Since accounting for logistics and production time is also necessary, it is necessary to find a tool that ensures that all indicators are taken into account while observing the target function.

However, the main requirement for the algorithm remains simplicity, flexibility and speed of operation. It is assumed that there are multiple launches of the algorithm when new orders are received for recalculation of the production program, which necessitates a high speed of the algorithm. Moreover, the process of forming the cutting cards takes place in a third-party program. This means that the algorithm must meet the requirements for flexibility and ease of integration with third-party services. Consider the existing approaches to optimizing production planning in terms of the following requirements:

- the search for a quasi-optimal solution, that ensures that all indicators are taken into account while observing the target function;
- high-speed operation and low consumption of resources;
- simplicity, flexibility and reliability of solutions as the need to accelerate technology deployments becomes even more critical;
- the ability to integrate with third-party information solutions.

As part of the analysis of existing solutions, the following tools were identified: mathematical techniques, heuristics techniques, simulation, and Gas (Guzman et al., 2021; Skornyakova et al., 2018; Fahimnia et al., 2013; Yang et al., 2016; Berbić et al., 2022; Chu et al., 2022).

Mathematical programming models such as Linear Programming models, Mixed Integer Programming models, and Lagrangian Relaxation models have been demonstrated to be useful analytical tools in optimizing decision-making problems. However, there are difficulties in using these methods in the framework of software implementation of optimization of production planning. Firstly, it is difficult to interpret the production planning process in the form of mathematical equations, since the increase in the number of variables and the introduction of new restrictions exponentially complicate mathematical algorithms. Secondly, the software processing of complex equations requires significant time and computer memory, which also affects the economic costs for the entire project (Fahimnia et al., 2013). The speed of calculation algorithms is a key requirement for the system, since it is assumed that the production program will be recalculated when new orders are received (Guzman et al., 2021; Skornyakova et al., 2018).

Simulation modeling allows you to simulate the behavior of a real system, which makes it possible to study the behavior of the system in the context of various external and internal changes (Yang et al., 2016). However, simulation modeling is not a tool for finding the optimal solution or program and is used for processes involving managerial influence on or within the system. In this case, simulation modeling is used to monitor the system as part of the diagnosis of problems and forecasts of its development depending on various changes. Heuristics techniques are also not used to find an optimal solution, therefore they cannot be considered for solving this problem (Fahimnia et al., 2013).

Genetic algorithms are principally able to search optimal solution because of robustness, searching flexibility and their evolutionary nature. Genetic algorithms produce a large population of solutions, for each of which the evaluation of the fitness function is sought (Berbić et al., 2022). The advantages of using GAs are their reliability, search flexibility and evolutionary nature. GAs is capable of searching in large, complex and unpredictable systems, which makes it easier to find the optimal solution. Thus, the choice of the optimal solution occurs by convergence of functions with the number of evolutions. There are, however, a number of challenges when designing a customized genetic algorithms procedure to solve a certain planning problem.

The first problem is to form the chromosome structure which accordingly affects the whole GA procedure (Chu et al., 2022). To solve different problems from various complexity levels it is necessary to make different chromosome representations. This means that Gas cannot be used to solve problems of different levels of complexity. The second problem is to make customized genetic operators to perform the matting process on the chromosomes. It turns out that there is an additional task of developing a mechanism for processing constraints. To solve the practical problem of production planning, it is proposed to find a simpler solution that will provide greater flexibility for improvements and expanding the capabilities of the algorithm.

The method of successive approximations allows you to get a solution with a certain accuracy in the form of a limit of consecutive iterations. In practice, this means that with each iteration, different constraints will be taken into account sequentially without departing from the objective function. Thus, by choosing the simplest method of optimization for one objective function, it becomes possible to take into account other features of the process and find a quasi-optimal solution a quasi-optimal solution that is as close as possible to the optimal one.

3 Results

Based on the results of the interview and the analysis of existing solutions for optimizing the planning of production processes, the following functional requirements for the algorithm were developed:

- calculation of production capacities taking into account a separate machine, a group of interchangeable machines, a workshop and a group of workshops;
- distribution of the production program according to the date of need by periods;
- checking the available production capacity before distributing the demand plan;
- recalculation of the production program when new orders are received;
- integration of the production program from several nearby production sites;
- distribution of the production program taking into account the characteristics of the machines;
- flexible settings of the reserve of production capacities in the context of a separate machine, a group of interchangeable machines, a workshop and a group of workshops;
- consolidation of the production program by the beginning of the period to ensure the reserve of production capacity for emergency orders.

Since the distribution of parts by cutting cards is the task of a third-party program, it is necessary to configure data exchange between the algorithm blocks and the program. The high-level algorithm of the program is presented in Fig. 1.

Figure 1 shows a simplified scheme of the algorithm. It is important to take into account that the input of information occurs sequentially in two blocks:

- production capacities;
- orders and production sites.

The information input block is divided into two branches, since joint data processing occurs only at the step of distribution of cutting lists according to production capacities.

It is important to note that the algorithm does not include choosing of the branches, because it is fully automated and continuous. All logical operators are designed to visualize internal checks in the distribution process of cutting lists according to production capacities.

Thus, the program receives data on production sites, orders and available production capacities. At the output of the program, the user receives a production program taking into account the distribution of cutting cards by production capacity in the context of a separate machine, a group of exchangeable machines, a workshop and a group of workshops.

Since the planning horizon is always a year, two types of periods are available within the calculation of the production program: a month and a quarter. The longer the period, the more parts fall into a single order and the greater the probability of obtaining a high material utilization rate. However, the probability of overhead costs for the storage of finished products increases.

The peculiarity of the distribution of the program is in compaction for the nearest period. For example, in the user set the period – month. Then, the calculation starts from December. The cutting cards with the material utilization factor are unformed and the details are sent to the previous period. Thus, the guarantee of manufacturing parts on

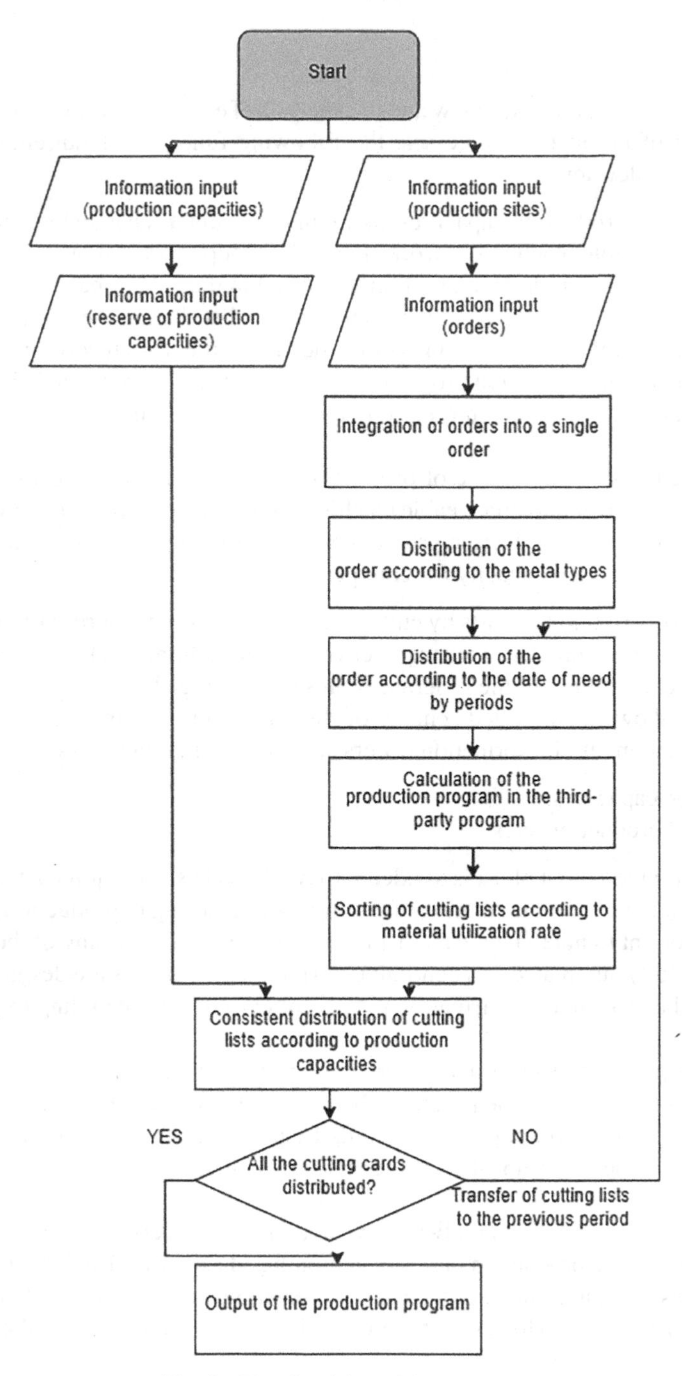

Fig. 1. The algorithm of the program.

time is provided. In the received production program, the earliest period will be filled as much as possible, providing an additional opportunity to recalculate the production program when urgent orders appear.

The program algorithm has a reserve of production capacity for emergency orders. The reserve data is filled in by the user. When a new order is received, the calculation of the production program is either started completely anew for all periods, or the program is recalculated for a period with the use of reserve capacities taken into account. The choice is up to the user.

It is also worth noting the peculiarities of the distribution of cutting cards by production capacity. The distribution takes place according to the characteristics of the material and on the basis of a larger mass. This means that the cutting card is sent to the site where most of the parts from the cutting card must be made according to the order. The developed algorithm is implemented in the program code and tested on various plans for the needs of the enterprise.

As part of the evaluation of the results of the algorithm, we will conduct a series of experiments:

- calculation without optimization – without taking into account the consolidation by periods and production sites;
- calculation with consolidation by periods;
- calculation with consolidation by periods and production sites.

The calculation was carried out for two production sites on 3 different plans. Each of the plans differs in the number of parts, the size of the parts and the corresponding amount of cutting lists.

Consider the calculation without optimization (Table 1). In order to maintain data confidentiality, all quantitative data are proportionally normalized.

Table 1. Results without optimization

	Production site 1		Production site 2	
	Number of parts	Material utilization rate (%)	Number of parts	Material utilization rate (%)
Plan 1	174876	80,9	207	70,6
Plan 2	51895	74,2	406	64,8
Plan 3	48718	59,2	4975	67,3

The calculation results reflect the real situation in production without taking into account the possibility of consolidation by periods and production sites.

Consider the calculation with consolidation by periods (Table 2).

As can be seen from the reference tables consolidation by periods leads to an increase for all production sites and for all plans. It is worth noting that an increase of more than 0.2% can already be considered cost-effective for the company due to the high cost of the material.

Table 2. Consolidation by periods

	Production site 1		Production site 2	
	Number of parts	Material utilization rate (%)	Number of parts	Material utilization rate (%)
Plan 1	174876	82,4	201	72,7
Plan 2	51895	74,7	404	65,2
Plan 3	48718	59,8	4644	72,1

Table 3. Consolidation by periods and production sites

	Consolidated calculation	
	The total number of parts	Material utilization rate (%)
Plan 1	543777	79,1
Plan 2	221548	72,0
Plan 3	206654	73,7

Consider the calculation with consolidation by periods and production sites (Table 3).

To compare the results, it is necessary to give the values of test 1 and test 2 to weighted average. Thus, we will be able to analyze the dynamics of changes in the indicator of material utilization rate for two sites at once.

Consider the results for each of the plans, taking into account the number of cutting cards (Tables 4, 5 and 6), the values of material utilization rate are calculated as weighted average for two sites.

Table 4. Summary results of the plan 1

	The total number of parts	The total number of cutting cards	Material utilization rate (%)
Test 1	543777	536	76,9
Test 2	543777	525	78,7
Test 3	543777	522	79,1

The results according to plan 1 showed an increase of about 2.2%, which allows to consider the distribution algorithm cost-effective.

Consider the results for Plan 2 (Table 5).

The results according to plan 2 showed an increase of about 0.7%, which allows us to consider the distribution algorithm cost-effective.

Consider the results for Plan 3 (Table 6).

Table 5. Summary results of the plan 2

	The total number of parts	The total number of cutting cards	Material utilization rate (%)
Test 1	221548	1311	71,3
Test 2	221548	1303	71,8
Test 3	221548	1298	72,0

Table 6. Summary results of the plan 3

	The total number of parts	The total number of cutting cards	Material utilization rate (%)
Test 1	206654	6323	65,6
Test 2	206654	5980	69,3
Test 3	206654	5625	73,7

The results according to plan 2 showed an increase of about 8%, which allows us to consider the distribution algorithm cost-effective.

It is worth noting that the size of the increase in the material utilization rate depends on the size of the parts. So the program with the most dimensional details gave the lowest increase.

Depending on the number of parts in terms of demand, the material utilization rate increased by 1–8%.

4 Conclusion

As part of this work, the task of using digital technologies to automate production planning was solved. In the course of the work, the features of the production process of the enterprise were analyzed and the functional requirements for the optimization algorithm of production planning were determined. The method of successive approximations was chosen as the main optimization tool, since it allows taking into account the existing production constraints without deviating from the objective function to increase the material utilization factor. The result of the work is an algorithm that optimizes production planning at a discrete machine-building enterprise. This algorithm allows you to combine plans for the needs of parts from several production sites, which leads to a more resource-intensive distribution of parts on cutting maps. The resulting cutting cards are distributed according to the production capacity of the enterprise, taking into account the processing characteristics of a particular material. The developed algorithm is implemented in the code and successfully tested on the plans of the enterprise's needs.

Acknowledgements. The research is funded by the Ministry of Science and Higher Education of the Russian Federation (contract No. 075-03-2022-010 dated 14.01.2022).

References

Alcácer, V., Cruz-Machado, V.: Scanning the industry 4.0: A literature review on technologies for manufacturing systems. Eng. Sci. Technol. Int. J. **22**(3), 899–919 (2019)

Athar, A.K., Janos, A.: Simulation of sustainable manufacturing solutions: tools for enabling circular economy. Sustainability **14**(15), 1–40 (2022)

Berbić, J., Ocvirk, E., Gilja, G.: Optimization of supervised learning models for modeling of mean monthly flows. Neural Comput. Appl. **34**(20), 17877–17904 (2022)

Chen, B., et al.: Smart factory of industry 4.0: Key technologies, application case, and challenges. IEEE Access **6**, 6505–6519 (2017)

Chu, J., et al.: A life cycle oriented multi-objective optimal maintenance of water distribution: model and application. Water Resour. Manage **36**(11), 4161–4182 (2022)

Dobrinskaya, D.E.: What is a digital society? Sociology of Science and Technology **12** (2021)

Fahimnia, B., Farahani, R.Z., Marian, R., Luong, L.: A review and critique on integrated production-distribution planning models and techniques **32**(1), 1–19 (2013)

Guzman, E., Andres, B., Poler, R.: Models and algorithms for production planning, scheduling, and sequencing problems: A holistic framework and a systematic review. J. Ind. Inf. Integr. **27**, 100287 (2021)

Kamble, S.S., Gunasekaran, A., Gawankar, S.A.: Sustainable Industry 4.0 framework: A systematic literature review identifying the current trends and future perspectives. Process safety and environmental protection **117**, 408–425 (2018)

Kritzinger, W., et al.: Digital twin in manufacturing: a categorical literature review and classification. IFAC-PapersOnLine **51**(11), 1016–1022 (2018)

Kuprijanov, Yu.V.: The formation of a model for integrated production planning. Bulletin of the Research Center of Corporate Law, Management and Venture Capital of Syktyvkar State University **1**, 35–44 (2018)

Qi, Q., Tao, F.: Digital twin and big data towards smart manufacturing and industry 4.0: 360 degree comparison. IEEE Access **6**, 3585–3593 (2018)

Ricciato, F., Wirthmann, A., Giannakouris, K., Reis And, F., Skaliotis, M.: Trusted smart statistics: Motivations and principles. Stat. J. IAOS **35**, 589–603 (2019). https://doi.org/10.3233/SJI-190584

Skornyakova, E.A., Vasyukov, V.M., Sulaberidze, V.: Algorithmisation methods for scheduling in high-performance assembly manufacturing. Bulletin of the Concern VKO Almaz-Antey **4**(27), 15–22 (2018)

Smirnov, A.V.: Digital Society: Theoretical model and Russian Reality. Monit. Pub. Opin. Econo. Soc. Chan. **1**, 129–153 (2021)

Uhlemann, T.H.-J., Lehmann, C., Steinhilper, R.: The digital twin: realizing the cyber-physical production system for industry 4.0. In: Umeda, Y., Kondoh, S., Takata, S. (eds.) 24th CIRP Conference on Life Cycle Engineering, CIRP LCE 2017, vol. 61, pp. 335–340. Procedia CIRP (2017)

Yang, Y., et al.: A parallel decomposition method for nonconvex stochastic multi-agent optimization problems. IEEE Trans. Signal Process. **64**(11), 2949–2964 (2016)

Digital Transformation of Enterprises Based on Analysis and Management Tools: Practical-Focused Research

Conceptual Model for Assessing the Formation of Universal Competencies in the Implementation of End-To-End Project Activities

Anton N. Ambrajei, Tatiana A. Its, Sergey G. Redko(✉), Alla V. Surina, and Inna A. Seledtsova

Peter the Great St. Petersburg Politechnic University, St. Petersburg, Russia
{ambrajei,its_ta,redko_sg,surina_av,seledtsova_ia}@spbstu.ru

Abstract. Modern educational standards are focused on the formation of universal competencies among students, and the project format is almost the only way to form the necessary flexible competencies. The ability to conduct project activities becomes one of the main meta-competencies. The most important element of the new edition of the educational policy of Peter the Great St. Petersburg Polytechnic University, published in 2021, is the concept of end-to-end project activities and the ecosystem of project activities. The implemented approach imposes appropriate requirements on the quality of graduate training, which, in turn, requires an appropriate system for measuring and evaluating the level of formation of universal competencies. The creation of such a system is an urgent problem now. The paper proposes a conceptual model for assessing the formation of universal competencies in the implementation of end-to-end project activities. Given the complexity of the internal structure of universal competence and the fact that the process of their formation is nonlinear, it is proposed to use "learning curves" for evaluation. For this purpose, the logistic form of the curve is considered. The concept of "project points" is introduced. A formal procedure for calculating project points is proposed, taking into account various characteristics of end-to-end project activities. The proposed approach to the assessment of universal competencies was tested at SPBPU for undergraduate courses. It is planned to test it as part of end-to-end project activities for undergraduate and graduate students in the framework of the Priority 2030 program.

Keywords: End-to-end Project Activities · Competencies · Universal Competencies

1 Introduction

The world is changing narrowly specialized knowledge and experience, are no longer as valuable in the eyes of employers as they used to be, if they are not backed up by universal skills. Mastering flexible skills allows an employee not only to increase the efficiency of work in his industry but also to remain in demand when moving between

V. Taratukhin et al. (Eds.): ICID 2022, CCIS 1767, pp. 73–82, 2023.
https://doi.org/10.1007/978-3-031-32092-7_7

industries. This is because professional knowledge becomes obsolete catastrophically quickly and it is easier to quickly train an employee with suitable flexible skills than to retrain a professional who does not have the necessary competencies [9].

Education should be reoriented to the post-industrial scientific paradigm and the realities of the XXI century; to interdisciplinary training and creative pedagogy; to develop the ability to find original effective solutions in non-standard situations, to work effectively in conditions of uncertainty; to perceive and implement innovations in all spheres of social production. In this regard, a number of main trends in the training of future specialists can be identified [10, 12]:

- formation of interdisciplinary skills;
- continuity of education (long-life-learning);
- practice-oriented education and involvement of representatives of industries in the educational process;
- digitalization of educational content and the educational process.

Modern educational standards are focused on the formation of universal competencies among students, and the project format is just about the only way to form flexible competencies that are so necessary for students. The ability to conduct project activities becomes one of the main meta-competences. There is currently no single format for organizing project activities at a university, so each educational institution develops its learning algorithm [13, 14].

In 2021, a new edition of the educational policy of Peter the Great St. Petersburg Polytechnic University (hereinafter referred to as SPbPU) [15] was released, one of the trends here is the concept of end-to-end project activity. Starting in 2017, SPbPU has introduced a mandatory basic discipline "Fundamentals of Project Activities", which is implemented for students of all areas of study. During this time, the university has created and successfully implemented a holistic system of involving students in project activities. It includes not only a hybrid course that is unique in structure, pedagogical design and architecture, but also a system of mentoring and involving external companies as customers of student projects.

During one semester, second-year students (4,500 people) go from choosing a project idea and finding a solution to receiving a product and presenting it to the customer, mentors and experts of the course "Fundamentals of Project Activity". Each launch is 650+ teams and about 100 mentors accompanying projects. The solution required special methodological, pedagogical and organizational solutions and, in fact, the creation of a new format of a mass course, which had not been conducted at the university up to this point. Since project training is impossible without the practical application of the theoretical knowledge obtained, the classic solution in the form of an online course could not ensure that students receive the necessary competencies. On the other hand, the organization of mass implementation of students' projects in the format of an individual or team project, as it is, for example, organized at the Moscow Polytechnic University is quite expensive.

Being in the fork between mass and individual training, the team turned to a hybrid format. The course consists of two blocks – theoretical and practical. Theory is a well-developed online course with an extensive block of onboarding, theoretical lectures, methodological guidelines and tests. It is characteristic that the introductory part of the course is onboarding, which allows you to combine both offline and online activities. Students gain applied project work skills by working in teams on real projects under the guidance of mentors. The main principle of this format is training through practice, aimed at implementing projects with specific customers. The results obtained:

For students: the development of soft skills, adequate self-esteem, increased motivation for project activities, participation in real projects of leading enterprises, with the possibility of further employment.

For the university: the experience of mass training of project activities was gained, teachers-mentors were trained; a data bank on the leaders of project activities (students and teaching staff) was obtained; networking with other universities was implemented; an increase in the number of integrated interdisciplinary projects carried out by university staff; new forms of interaction with partner enterprises.

In the new educational policy, the concept of end-to-end project activity is normatively fixed, which includes an obligatory part – the course "Fundamentals of Project Activity", as well as a variable component, which includes disciplines that provide for the implementation of term papers and course projects implemented in the form of a project, the discipline of the mobility module "Tools and technologies of project activities", research, final qualifying work, undergraduate practice (if there is a component of project activity). As a result, an ecosystem of project activities is being formed at the university, capable of quickly adapting and responding to the needs of students, teachers, industrial partners, companies and the transformation of society as a whole.

The mission of this ecosystem is the mass implementation of integrated project activities using STEM/STEM technologies. As a result, such an environment will be created for students where theoretical knowledge provided by the university is combined with research and practical tasks from real companies to develop students' skills that increase their competitiveness in the global labor market after completing their studies at the university.

The approach being implemented imposes appropriate requirements on the quality of graduate training, which, in turn, requires an appropriate system for measuring and evaluating the level of formation of universal competencies and meta-competences. The creation of such a system is an urgent problem now. The purpose of the work is to propose a conceptual model for assessing the formation of universal competencies in the implementation of cross-cutting project activities.

2 Methods and Materials

From the point of view of measurement and evaluation, universal competencies are complex objects of a latent nature. The process of forming universal competencies is a continuous process that is extended over time. Today, there are two complementary approaches to assessing competencies. In the first approach, based on the decomposition of competence into indicators and descriptors (knowledge – skills – possessions), a matrix of distribution of descriptors by discipline elements, including types of assessment tools used, is compiled. This approach works well if we are dealing with one discipline that forms hard or professional competencies. An alternative is an approach in which a portfolio is assessed based on a student's individual achievements.

However, if we are trying to evaluate universal competencies, then the level of competence formation will depend on many factors. To date, there is no clear understanding of how to assess the competence formed by a pool of disciplines in dynamics (throughout the entire training, several semesters). The article [5] describes in detail the problem of measuring universal competencies – it is indicated that the very concept of "universal" implies the presence/formation of these competencies in all modules of educational programs and in various activities. Since universal competencies are of an over subject nature, their formation is carried out within the framework of various forms of organization of the educational process throughout the entire period of study. Therefore, to evaluate them, it is necessary to use multi-stage multidisciplinary meters. In this case, the transition from decomposition to integration of educational results will be justified. This means a transition to interdisciplinary character, meta-subjectivity and the cross-cutting nature of their formation.

3 Results

The competency assessment tools are considered in sufficient detail in [2], and more specific questions about measuring competencies are discussed in [1] and [11]. As a comprehensive assessment of the level of competence formation, these works propose to use additive convolution. The use of additive convolution in the evaluation is justified only in the case when the results of the mandatory part, implemented within the framework of the discipline "Fundamentals of Project Activities", are evaluated.

It is proposed to use the following methodology to evaluate the results of training in the discipline "Fundamentals of Project Activities" [10]. First, in accordance with the competence matrix, indicators of the achievements of universal competencies formed by the student during the development of this discipline are highlighted. Secondly, the obtained indicators, if necessary, are grouped so that the total number of evaluation indicators (indicators of achievement) is optimal for visual representation (for example, $10 \leq n \leq 14$). Thirdly, the composition of control and measuring materials and means of assessing competencies for each indicator of achievement of learning outcomes in the discipline is determined. This approach is used quite widely in SPbPU, but a characteristic feature of the proposed methodology for evaluating students' results in the discipline "Fundamentals of Project Activities" is that:

- weighting coefficients are determined taking into account the share of participation of each evaluation tool in the formation of indicators and competence;
- both individual students and teams are evaluated, both the mentor teacher and the head of the student project team contribute to the evaluation. The practical part of the course is based on the teamwork of students and, as a result, each student receives two grades: an assessment of the work of the team and an individual assessment of the student. During the training, the team performs group tasks (templates and presentations). The evaluation of each template/presentation contributes to the evaluation of several indicators, which in turn contribute to the development of several competencies with a certain weighting factor.

Given the complexity of the internal structure of universal competence and the fact that the process of their formation is non-linear, it is possible to use learning curves for evaluation [6, 7]. It is proposed to consider two main processes for obtaining useful information in the course of training: iterative and logistical. The iterative process assumes that the speed of assimilation of information is proportional to the speed of its receipt and decreases with the growth of the already learned [8]. This process is described by an exponential curve. In contrast to the exponential, the logistic learning curve is characterized by an initial flat area of information accumulation, after which there is a sharp increase in the rate of assimilation of information or the formation of competence.

To assess the learning outcomes in the discipline of general education, it is proposed to use the following methodology [3, 4]: decomposition of universal competencies formed by the discipline into indicators of achievement of relevant competencies; considering the different contributions of the theoretical and practical parts of the course to the development of various indicators. Table 1 presents a matrix of universal competencies built in accordance with this methodology.

Table 1. Competence matrix for UC-2 and UC-3

Competence category	Code	Name of the competence	Indicators	Theoretical part	Contribution to competence	Practical individual assessment	Contribution to competence	Final assessment of the level of competence formation
	Yj		p	*Tkpj*	*cpj*	*Nkpj*	*dpj*	*Yjk*
Development and implementation of projects	UC-2	Is able to determine the range of tasks within the set goal and choose the best ways to solve them, based on existing legal norms and available resources and limitations	Formulation of tasks that ensure the achievement of the goal		0,05		0,15	*Y2k*
			Information search		0,02		0,08	
			Information analysis		0,02		0,08	
			A systematic approach to solving the tasks set		0,02		0,08	
			Designing a solution to a specific problem		0,05		0,15	
			Solving specific tasks of the project of the declared quality and in a set time		0,05		0,15	
			Presentation of results		0,02		0,08	
Teamwork and leadership	UC-3	He is able to carry out social interaction and realize his role in the team	Leadership		0,01		0,04	*Y3k*
			Communication		0,05		0,25	

(*continued*)

Table 1. *(continued)*

Competence category	Code	Name of the competence	Indicators	Theoretical part	Contribution to competence	Practical individual assessment	Contribution to competence	Final assessment of the level of competence formation
	Yj		p	*Tkpj*	*cpj*	*Nkpj*	*dpj*	*Yjk*
			Personal contribution (Adequate personal assessment)		0,02		0,08	
			Information and communication technologies		0,05		0,15	
			Teamwork		0,05		0,25	
			Activity		0,01		0,04	

In general, the level of competence formation in the course of mastering a particular discipline is determined by the formula (1):

$$Y_{jk} = \sum_{p=1}^{P} (N_{kpj} d_{pj} + T_{kpj} c_{pj}) \tag{1}$$

where:

Y_{jk} – the level of mastery of a specific student j competence;
N_{kpj} – individual assessment for the practice of a particular student according to the *p* indicator (indicator) *j* of competence;
T_{kpj} – individual assessment of each student k, received for work on the theoretical course on the p indicator (indicator) *j* of competence;
d_{pj} and c_{pj} – coefficients showing the contribution of individual grades of each student *k*, received for practice and for work on the theoretical course (respectively) for *p* indicator (indicator) *j* of competence.

The strength of the proposed assessment of the results of students' achievements in the discipline of general education is: taking into account the contribution of the assessment tools used, templates and test results (of various types) to the development of indicators by assigning weighting factors, taking into account the share of participation of each assessment tool in the formation of indicators and competence; both the student and the team are assessed, and both the teacher-mentor and the head of the student project team contribute to the assessment.

Thus, the individual assessment of student k by the indicator *p* (N_{kp}) takes into account the student's "personal contribution" to the work on the template/presentation. The individual assessment of each student is found as the product of team scores and "personal contribution" (see (2)):

$$N_{kp} = \sum_{i=1}^{l} R_i a_{ik} b_{ip} \tag{2}$$

where:

R_i – a team assessment for template *i*, the completion of which contributes to the formation of the competence for which assessment is carried out N_{kp}.
$R_i a_{ik}$ – individual assessment of each student *k*, received for work on template *i*;
a_{ik} – coefficient showing the contribution of a particular student to the grade for the completed template *i*. The coefficient takes a value in the range from 0 to 1 in increments of 0.1, depending on the student's contribution. "Personal contribution" is evaluated by the head of the project team; is a weighting coefficient showing the contribution of a specific template (estimation) to the formation of the indicator p.

The evaluation of the work of the team s by the indicator p (M_{sp}) is determined as follows (3):

$$M_{sp} = \sum_{i=1}^{I} R_i b_{ip} \quad (3)$$

where:

– team score for template i, i = 1, …, 11 (the number of completed templates);
$R_i b_{ip}$ – contribution to a specific template (estimation) R_i to the formation of the indicator p;
– coefficient (factor).

Note that a team assessment is an assessment that goes only to the student's portfolio.

A graphical representation of the learning outcome of each student (individual and team) is shown in Fig. 1.

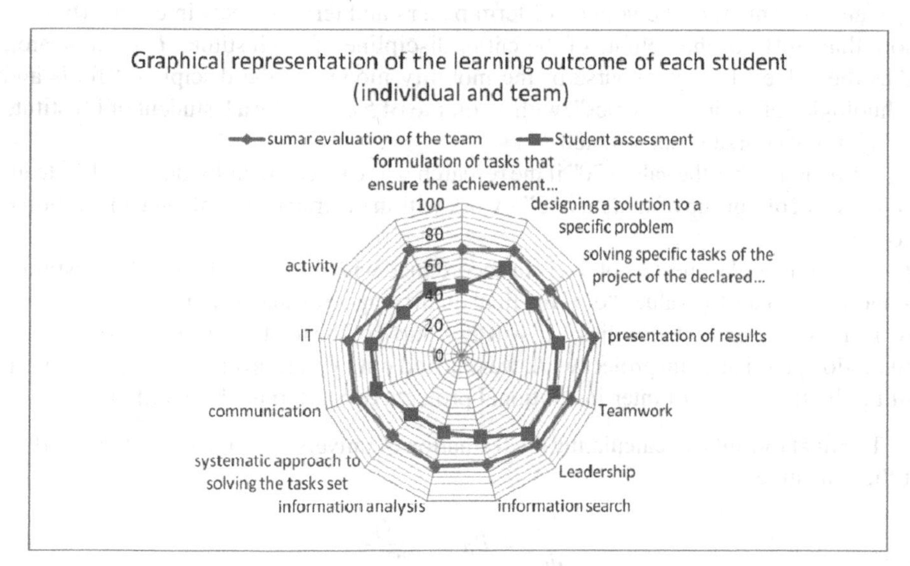

Fig. 1. Diagram (visualization) of assessment (team and individual) by a group of indicators (indicators of competence achievement)

In our case, when obtaining a comprehensive assessment of discipline competence, we divide the results of mastering conditionally into two components (practical and theoretical) because it is not possible to separate the skills and practical skills in this case.

The proposed approach for evaluating end-to-end project activities on the example of SPbPU assumes:

- Recording of all project activity on the portal, formation of a digital trace: an element of the curriculum, connection with project activity, information about the project, progress, student assessment, accumulation of "project points" of the student.
- The result of individual project activities – each student is required to score a certain number of "project points".

The level of competence formation during the learning process is measured in project points P_l.

$$P_{lk} = \sum_{i=m}^{M_l} R_{il} f_{ikl} + 5e_{kl} + \sum_{q=1}^{Q_l} 3h_{qkl} + 3s \tag{4}$$

$l = 1, \ldots,$ L и $k = 1, \ldots,$ K

Here, P_{lk} – the number of project points scored by k student of l institute; M – the total number of professional disciplines with term papers and term projects in the curriculum k of a student of l institute, $i = 1, \ldots,$ M.

- indicator, takes the values "0" if the work was carried out by k student of the l institute individually, «1» – team work of a group of students, «2» – in inter-institution projects. R_{il} – an assessment of the volume of term papers and term projects in credits (but not more than 80%) of the volume of the entire discipline of the institute. E_{kl} – indicator, takes the value "1" if the course of the mobility module is the discipline "Tools and technologies of project activities", with a volume of 5 credits by a k student of l Institute or "0", if the course is not selected.
h_{qkl}- indicator, takes the values "0" if the research was carried out individually, "1" – team project work of a group of students, "2" – work within the framework of inter-institutional projects.
Q_l – the number of semesters of R&D in accordance with the curriculum of the l institute. s – indicator, takes the values "0" if the final qualifying work and pre-graduation practice are not related to project activities, "1" – in the course of the final qualifying work and pre-graduation practice, team project work was carried out by a group of students, "2" – work within the framework of inter-institutional projects or Startup is like a diploma.

The final formula for calculating the formation of universal competence j for k student of the l institute:

$$\frac{dY_{jk}}{dt} = P_{lk} \frac{(1 - Y_{jk})}{P_l^{\max}} \tag{5}$$

$l = 1, \ldots, L$ и $k = 1, \ldots, K$

Where P_{lk}- number of project points scored by k students of the l institute; $P_l^{\max}$ is the maximum possible number of project scores of student l Institute; Y_{jk} the level of formation of universal competence j for k student of the l institute.

Thus, with the help of "learning curves" a relationship is established between students' assessments and the level of formation of individual components and parts of the declared competencies.

The proposed approach to the assessment of universal competencies was tested for junior bachelor's students at the FPD course, which has been implemented since 2017. It is currently planned to be tested as part of a pilot launch of cross-cutting project activities for undergraduate and graduate students under the Priority 2030 program.

4 Conclusions

An approach to assessing the formation of a student's universal competencies in the process of implementing cross-cutting project activities is proposed. The organization of end-to-end project activities involves fixing a digital trace – information about ongoing projects, progress, assessments of each student at various stages. This provides an opportunity to evaluate the level of both an individual student and a project group. The use of learning curves is proposed as the basis for the formation of an assessment, which allows, firstly, to consider the non-linearity of the process of forming competencies. Secondly, it reflects the dynamics of the process of formation of universal competencies, as a continuous process. To assess the level of competence formation during the learning process, the concept of "project points" is introduced. A formal procedure for calculating project scores is proposed, considering various characteristics of cross-cutting project activities. The proposed approach allows you to automate the process of quality control of student training.

At the moment, an automated system for monitoring and visualizing learning outcomes in disciplines included in the module of end-to-end project activities is being tested, where the collection of a "digital footprint" will be implemented and methodological recommendations for teachers and students will be prepared. Thus, the result of the research is the proposed comprehensive approach (procedure) for assessing the level of formation of students' competencies in the field of project activity, including a formalized (mathematical model) assessment of students' achievement of indicators of the assessed competencies.

The approbation was carried out when evaluating the results of students' training within the framework of the discipline "Fundamentals of Project Activity", which form two universal competencies ("meta-competencies") related to students' project activities. At this stage, the proposed model showed a high level of reliability of the results obtained. However, when evaluating other competencies, and primarily from the professional group, the proposed approach will require refinement in terms of taking into account the hypothesis of learning curves.

Conflict of Interest Statement. On behalf of all authors, the corresponding author states that there is no conflict of interest.

References

1. Devisilov, V.: Tools for competency assessment and diagnostics of knowledge on the example noxological competencies and disciplines. Stand. Monit. Educ. **1**, 3–12 (2011)
2. Efremova, N.: Competences in Education: Formation and Evaluation. National Education, Moscow (2012)
3. Its, T., Redko, S., Chenikova, A., Schepinin, V.: Formation of a competency-based model of learning outcomes in a particular discipline. In: Qualification Assessment System in the Development of University Education in Russia and Foreign Countries, pp. 60–63. Yekaterinburg (2018)
4. Its, T., Surina, A.: Evaluation of student learning outcomes in an electronic educational environment: on the example of the discipline "Fundamentals of project activities." Naukosphere **3**(1), 80–85 (2022)
5. Kazakova, E., Tarkhanova, I.: Evaluation of the universal competencies of students in the development of educational programs. Yaroslavl Pedagogical Bull. **5**, 127–135 (2018)
6. Novikov, D.: Patterns of Iterative Learning. Institute for Control Problems. Russian Academy of Sciences, Moscow (1998)
7. Ovchinnikov, A.: A model for the accumulation of acquired knowledge and acquired competencies based on learning curves. Modern High Technol. **5**, 50–53 (2017)
8. Ovchinnikov, A., Gitman, M.: Quality management of educational programs for the preparation of students with the use of corrective actions based on the negentropic approach. Appl. Math. Control (2), 97–108 (2018)
9. Ponomareva, O.: Formation of soft skills (soft skills) as a condition for the adaptation of the modern generation to the labor market. In: Current Problems of Social Professional and Economic Entry of Youth in the Regional Social and Production Environment 2018, pp. 29–33. Yekaterinburg (2018)
10. Redko, S., Tsvetkova, N., Seledtsova, I.: A systematic approach to training specialists for a new technological order. Lecture Notes Netw. Syst. **95**, 643–650 (2020)
11. Stolbova, I., Danilov, A.: Toolkit for evaluating the results of education with a competency-based approach. Educ. Standards Monit. **4**, 24–30 (2012)
12. Tukkel, I.: Tomorrow, the technological tomorrow, came yesterday. Innovations **11**, 3–5 (2017)
13. Basics of project activity. http://project.spbstu.ru. Last accessed 10 Jul 2022
14. Own university educational standards of SPbPU. https://www.spbstu.ru/sveden/eduStandarts. Last accessed 10 Jul 2022
15. Educational policy of FAGBOU HE "Peter the Great St. Petersburg Polytechnic University". https://www.spbstu.ru/education/general-information/educational-activities. Last accessed 10 Jul 2022

Ecosystem Courses as an Effective Way to Prepare the SAP Talent Pool

Anton N. Ambrajei[1], Nikita M. Golovin[1], Anna V. Valyukhova[1(✉)], Natalia A. Rybakova[1], and Yury V. Kupriyanov[2]

[1] Peter the Great St. Petersburg Polytechnic University, 29 Polytechnicheskaya Street, St. Petersburg 195251, Russia
avalyukhova@spbstu.ru

[2] SAP, 52/7 Kosmodamianskaya Embankment, Moscow 115054, Russia

Abstract. This work summarizes more than 10 years of the authors' experience in creating and delivering SAP courses for students and teachers of higher educational institutions. In this article, the authors consider the difficulties that universities face in preparing such specialists and propose a solution in the form of a multi-format hybrid course, as well as describe their experience in conducting such courses.

The article describes in detail the structure of the hybrid course, its participants and the forms of their interaction, various formats of the training, depending on the needs and situations. It also compares the two SAP Academies that took place in 2020 and 2021. Their differences are considered in terms of student recruitment, the onboarding process and progress monitoring, as well as how all these factors influenced learning outcomes.

This work can be useful both for specialists in online and blended learning and for representatives of large companies interested in competently preparing the future pool of talent for their ecosystem.

Keywords: ecosystem course · e-learning · hybrid course · multi-format learning environment · higher education · SAP · S/4HANA · MOOC

1 Introduction

Almost all major IT companies have their own programs for interaction with universities. Companies implement various approaches: the creation of joint training programs, from separate courses to full-fledged master's programs, the opening of joint laboratories or industrial chairs.

Each of the options has its own advantages and disadvantages. In general, universities are faced with the difficulty of training teachers for specific solutions, the gap between the theory and practice of applying solutions, and the lack of direct communication with employers.

SAP company in this respect implements one of the most complete and systematic approaches to learning. All educational materials are developed and updated centrally, and universities also get access to cloud IT systems to perform practical exercises in

V. Taratukhin et al. (Eds.): ICID 2022, CCIS 1767, pp. 83–95, 2023.
https://doi.org/10.1007/978-3-031-32092-7_8

accordance with these educational materials. Initially, materials and systems are available in English and German, adaptation and translation into other languages is carried out by local Academic Competence Centers in universities.

Thus, a ready-to-use educational product is obtained - a training system, theoretical materials and instructions for performing practice. All materials are designed as independent modules, which makes it possible to embed them as practical examples not only in IT courses, but also in courses on finance, logistics, production process management, etc.

However, this approach does not solve all the issues of interaction with employers and mass high-quality training, because not all universities have sufficient resources to train competent teachers [1].

Peter the Great Polytechnic University, together with the SAP University Alliances CIS, has proposed a fundamentally new approach to learning: ecosystem courses on the flagship SAP product, ERP system SAP S/4HANA.

2 The History of the Development

In 2008, the ACC SAP SPbPU became the main center for coordinating the teaching of SAP technologies in the CIS countries. The tasks of the center included training of teachers and students, developing a methodology for introducing SAP courses into the educational process, organizing and conducting certification courses. Also the area of responsibility of ACC SAP SPbPU included the creation of new and adaptation of the standard courses of the SAP University Alliances, preparation and delivery of SAP training systems to universities.

The SAP University Alliances program was launched in Germany in 1988, and more than 1,000 universities from around the world took part in it [2].

The goal of the SAP University Alliances program is to stimulate inter-university cooperation, as well as to unite the efforts of SAP and educational institutions around the world to improve the quality and level of education and, ultimately, the innovative component of the scientific, technical and industrial sector.

Curricula in the SAP University Alliances are developed centrally through the University Competence Centers (UCCs), and the academic centers are engaged in the localization and development of specialized topics. Access to the software is carried out through hosting centers, so universities do not have the difficulties associated with installing, configuring and maintaining their own training systems [3].

For the last 10 years, the center has conducted more than 80 training courses, and about 3,400 students and teachers took part in them. Of these, 30 courses were certification courses under the TERP10 "SAP ERP Integration of Business Processes" and TS410 "Integrated Business Processes in SAP S/4HANA" programs, and as a result, 272 people received professional certificates (Fig. 1).

The peak of certification came in 2012, when several leading universities joined the courses and organized certification centers. In general, the maximum courses were in 2019, the center opened an additional program on analytical solutions, including cloud ones.

Nevertheless, training until 2019 was carried out quite traditionally, in full-time or part-time format, only the number of participants and the location varied. The standard introductory course consisted of a 4–5 day's full-time part, which was conducted for teachers, and their subsequent independent work with the system using ready-made methodological materials.

Face-to-face teaching has an undeniable advantage because the knowledge is transferred "from hand to hand", there is an opportunity to provide personal support and, to some extent, inspire teachers to learn new complex technologies.

However, there were also a number of problems. Introductory training alone was clearly not enough, a combination of introductory training and a two-week certification course was recommended, but not so many teachers were ready for such an intensive training. Also, often teachers could not find enough time for continuous learning. For these reasons, there was a lack of highly qualified teachers in the field of SAP technologies. It was necessary to use new tools and approaches to learning.

At the beginning of 2019, the authors began the preparation of an online course on the Moodle platform: recording video lectures and video instructions, developing notes and guidelines for practice. Simultaneously with the reworking of the content, the study of the specifics of hybrid courses and the best world practices in the development of MOOCs was going on [4, 8, 9, 11]. After analyzing the existing practices for the implementation of online and blended learning and flipped classroom courses, the authors came to the conclusion that it was possible to create some universal teaching methodology that will combine the best qualities of the existing ones [7].

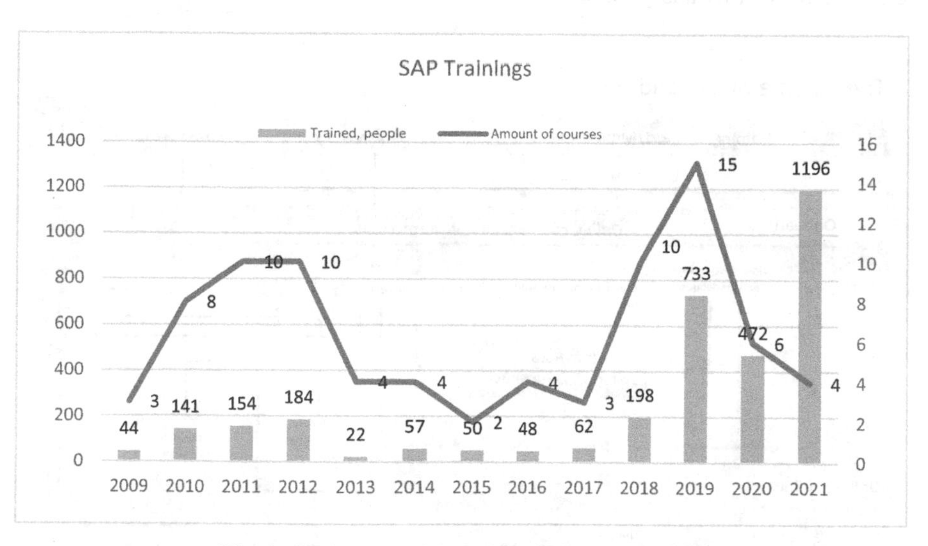

Fig. 1. Trainings conducted by SAP ACC SPbPU

3 Multi-Format Hybrid Courses

The main idea of such courses is that the teacher should have a set of modular materials and communication tools that he is able to quickly adapt (recompile) to the objectives of a particular course in specific conditions. Courses should be initially designed with this approach in mind [10, 14].

Let's see how the course format affects the composition and use of the main elements. We will consider an online course in a hybrid model, blended learning (intensive and scheduled classes, remote intensive).

Each of the formats has been used by teachers and can be considered quite well-tested.

The key element of the scheme shown in Fig. 2 is an online course implemented on LMS Moodle. Moodle is used by a large number of universities and is a standard platform for SPbPU, students are accustomed to working with this LMS, and it is understandable to most of the teachers. Moodle's technical capabilities make it easy to create and rebuild courses, quickly publish videos, and work with question banks. As a result, we came to a model where a master course exists and it is "cloned" for different purposes. Reconfiguring the course to the desired format takes about an hour.

The online course contains standard elements for this format: video lectures, abstracts, tests, guidelines for practice, as well as video instructions for execution of the case studies.

Despite the fact that the practice is performed in a separate system, the practice check is carried out through the LMS tools, which allows you to record an individual track of the course for further analysis.

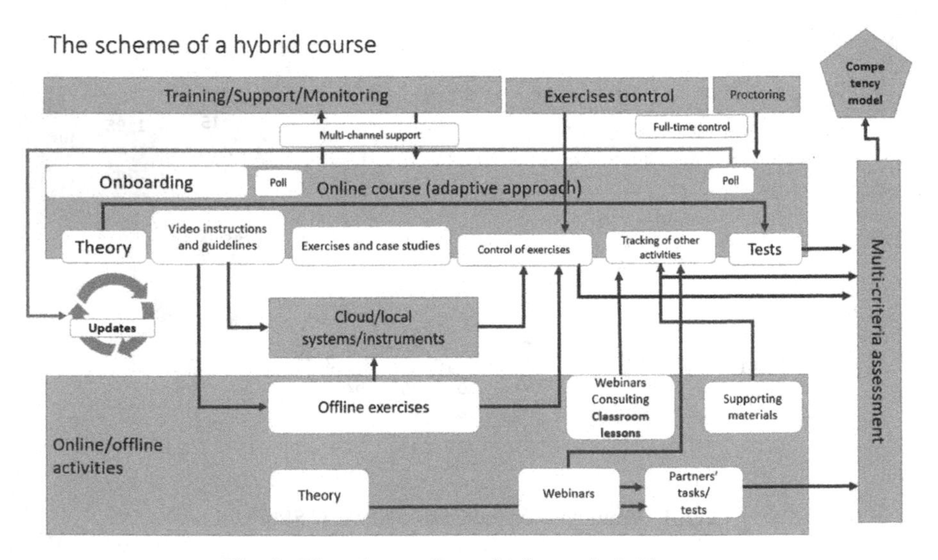

Fig. 2. The scheme of a multi-format hybrid course

The hands-on part of the course is implemented on the S/4HANA ERP system, which is deployed in the hosting center. This solution makes it easy to organize access to practical tasks from anywhere in the world, and the SAP FIORI web interface allows you to use only a browser for work. The SAP GUI interface is available as an option for students and they can run cases using it if they wish. In general, for such courses, cloud solutions are preferable to on-premise solutions.

Students work in S/4HANA on the data of a model company with customized and transparent processes. Clear processes and a well-described company structure help students understand their operations. All processes are end-to-end and are tied to the real roles of employees, performing the process, and the student goes through 5–8 different roles in one case study. Data sets for students are divided according to the principle of adding an individual three-digit number to all entities.

To automate the verification of user actions in the training system, there are special transactions that greatly simplify the search for errors and scoring.

The first launch of the hybrid online course was in September 2019, about 400 students from 60 + universities of the Russian Federation, the Republics of Belarus, Kazakhstan and Uzbekistan took part in it [6]. The results of the course are described in detail in the article [7].

The second launch on March 1, 2020 was more ambitious - already more than 660 students from 80 universities. It had two streams - an open recruitment and a target recruitment of the SAP personnel reserve from a limited list of universities, in this case, the responsibility for the contingent lay with the curators from the universities.

In total, 590 out of 1,170 participants completed the first version of the hybrid course. In addition to 2 online launches, there were three face-to-face courses and one remote intensive course. All formats performed quite well [12].

4 Format Features

Online course on the hybrid model. The course is held in a standard format "weekly" for 10–15 weeks, topics are opened sequentially, but it is possible, by completing additional optional tasks and tests, to open topics faster. This opportunity motivates listeners to better master the material. According to our estimates, about 20% of students use it.

Support for listeners is multi-channel: forums on each topic, a dedicated email address, weekly newsletters, video consultations via zoom, and for the remote intensive format and groups of up to 40 people - a WhatsApp group chat. Recently, forums have been abandoned, as it turned out that students practically do not communicate on the forum, but are waiting for answers from teachers. The effectiveness of the chat is also called into question, because instead of solving the case on their own, students begin to ask questions at each step.

With a large number of participants, the onboarding module is very important, and the authors of the course paid great attention to it. Such elements as checklists, knowledge bases (for example, on practice mistakes), video instructions, etc. were used. The main technical difficulty for the trainees was the need to work both on the online Moodle-based platform and in the ERP system at the same time. Therefore, detailed instructions and videos have been developed. High-quality onboarding allows you to remove a lot of questions. With a small number of listeners, an introductory webinar is usually sufficient.

Mixed format. It combines a full-time intensive course lasting 5 days or a regular course for students with classes in the classroom according to the schedule, and an online course as a tool for testing, fixing results and self-study and practice.

The remote intensive is distinguished by the replacement of the face-to-face part with a series of webinars and more intensive support, for 5 days the teacher is constantly in touch with the group.

5 Benefits of the Multi-Format Approach

The multi-format approach has a number of clear advantages:

- **Reuse of content.** Thanks to the modular approach - essentially micro-learning - different "learning blocks" can be easily combined into new courses.
- **Solving problems with absenteeism.** Since the bulk of the materials are available on the online course, students who missed classes will be able to fill in the gaps and even complete the practice on their own.
- **Available storage.** Students can download the materials they need at any time.
- **Testing platform**. Modern LMS systems are great for testing, and with proctoring tools you can get objective results when testing outside the classroom
- **Accumulation of material** - an online course is a good tool for teachers to collect and enrich their courses
- **Fast Change**. Modifying a course through an LMS is easy, although it does require some technical skill, and updated materials are immediately available to all course participants.
- **Recording the digital footprint** - you can analyze the activity of students, the results of tests and practical classes, even mark the attendance of face-to-face classes. This creates the basis for an objective assessment of the final grade.
- **Adaptability** - you can quickly adjust the course to changing conditions or audiences, there are no problems with adding content.
- **No problems with distance learning**. If you have a foundation in the form of an LMS course, then the transition to remote learning is painless, as was the case with all SAP ACC courses during COVID-19 restrictions.
- **Adding partner content**. If you need to connect an external expert to the course, you can do it directly through the LMS platform or upload a record.
- **Export of education**. Since we have a finished product, with the possibility of remote participants, it can be included in the educational process of other universities.

6 The Architecture of the SAP Academies

The 2020 and 2021 SAP Academies were the largest project that used the above approach. These were free and for the most part of the students not compulsory courses.

In total, three courses were held - a pilot and two full-fledged academies.

The enrollment in each academy was 1000 students from Uzbekistan, Kazakhstan, Azerbaijan and the Russian Federation. Students were selected on the basis of analysis of their learning and motivation profile according to the point-rating model. In addition, the audience included target groups from specific universities. To increase students' motivation, meetings were held at universities with representatives of SAP partners and customers before the start of the course.

The existing hybrid course on S/4HANA was expanded with two additional training circuits, and a number of organizational and career guidance activities were carried out within the course. Companies participating in the course also had the opportunity to prepare their assignments for students and evaluate them. Thus, students received marks in several categories: theory, practice, webinars from SAP and partners, and assignments from partners.

The first circuit is in-depth seminars on specific SAP modules and solutions, the second is the soft skills required for an SAP consultant.

Below are the results of student surveys: the first one was taken at the beginning of the course, and the second one in the middle. It can be seen that the profession of consultant remains a gray area for students, although understanding is improving (see Fig. 3).

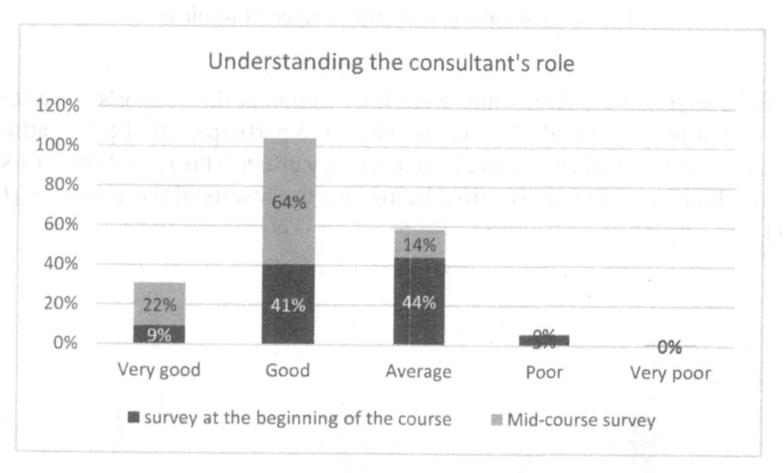

Fig. 3. Dynamics of understanding the role of the consultant

The large number of participants made it possible to organize career guidance events within the course, such as Career Day and Job Fair, which were very popular with students.

Additional materials were implemented in the form of webinars delivered by SAP experts and partners. The control of attendance and assimilation of materials was included in the general system of student assessment.

The average results of two surveys on the value of webinars show that students rate these sessions well (Fig. 4).

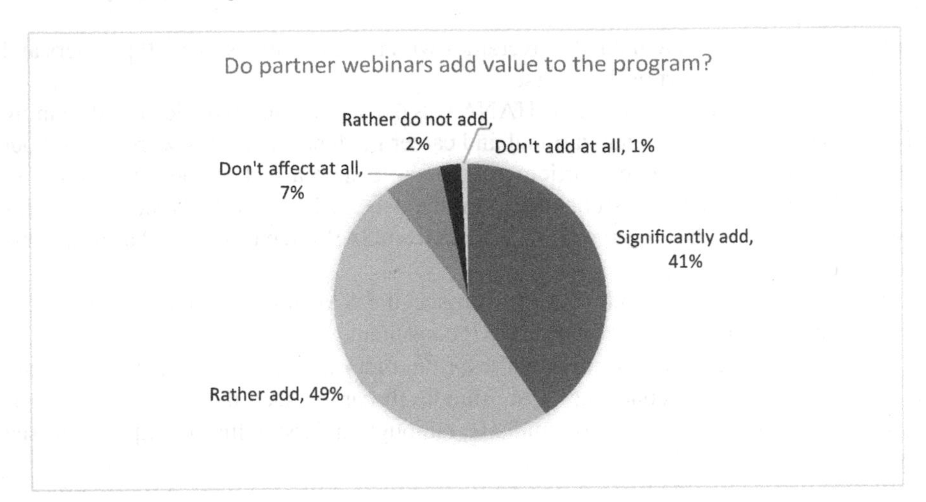

Fig. 4. Assessment of the importance of webinars

It can be said that the Academies were fully aimed at the vendor's ecosystem as a whole, and not just at individual products. The wide participation of SAP partners and customers made it possible to achieve a high synergy effect. This is confirmed by surveys that show a high degree of satisfaction. Below are the results of these surveys (Figs. 5, 6 and 7).

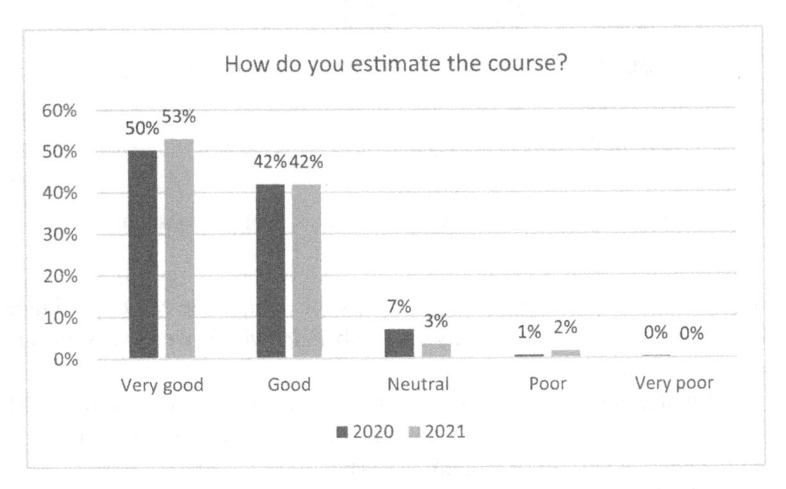

Fig. 5. Comparison of overall course grade across two Academies

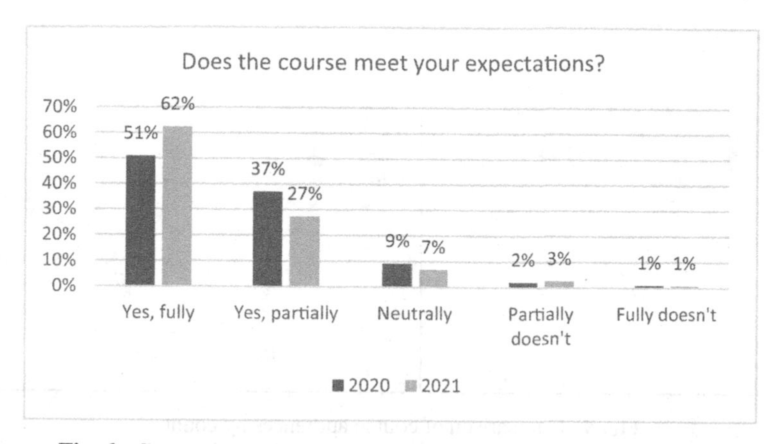

Fig. 6. Comparison of course compliance with trainees' expectations

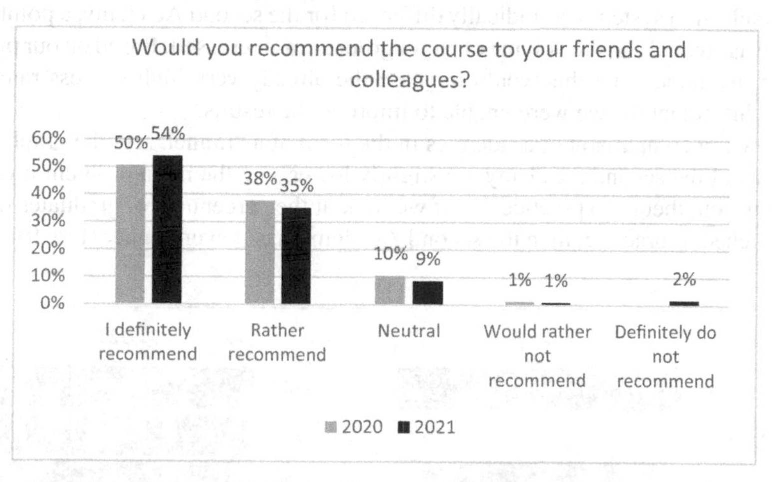

Fig. 7. Comparison of willingness to recommend the course

7 The Results of the SAP Academies

Both Academies had the same structure, but there were some differences. The main differences were in the composition of the participants, the structure of the webinars and how they were evaluated. In the first Academy, webinars were evaluated by the presence of listeners, and in the second, by mini tests. In addition, two more cases on system settings were added in the 2021 course.

The input number of participants for both courses was approximately the same - about a thousand students (Fig. 8).

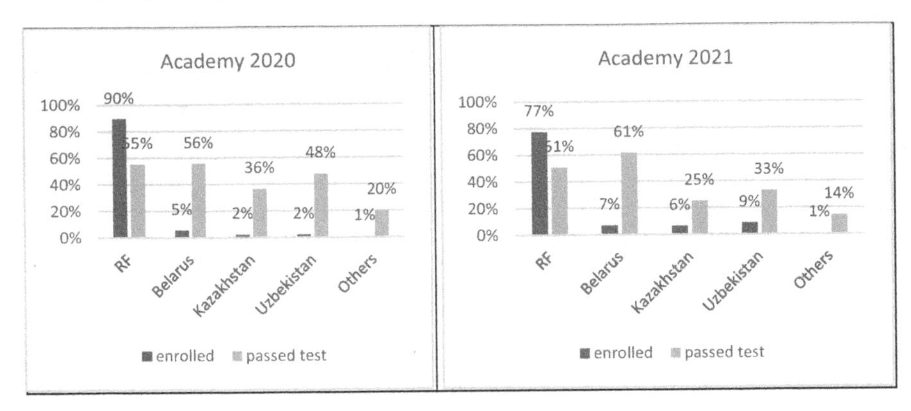

Fig. 8. Comparison of course audiences by country

The selection system was radically different: for the second Academy, a point-rating system was created and the volume of target groups was increased. Based on our previous research, we hoped that this would increase the already very high success rate of the course. Unfortunately, we were unable to improve the results.

Below is a comparison of academies in the form of a "funnel". As the graph shows, the results of the second Academy are slightly lower, and the main problem lays in the transition from theory to practice, but if we look at the percentage of graduates of those who switched to practice, then the second Academy has it even higher (Fig. 9).

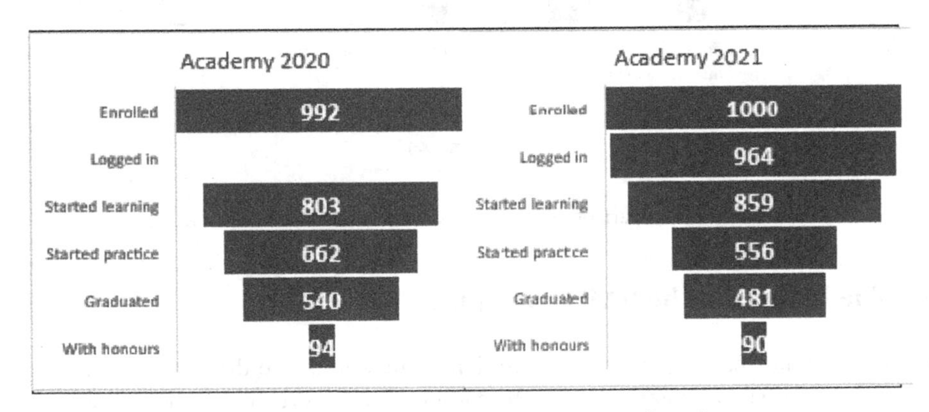

Fig. 9. Comparison of course results as a funnel

The behavioral characteristics of the participants in both courses are almost the same. Below is the distribution of online course attendance by participants and the number of actions on the learning platform in the Academies 2020 and 2021, respectively. As one can see on the graph, the students show approximately the same behavior with peak efforts during the onboarding phase and near deadlines (Fig. 10).

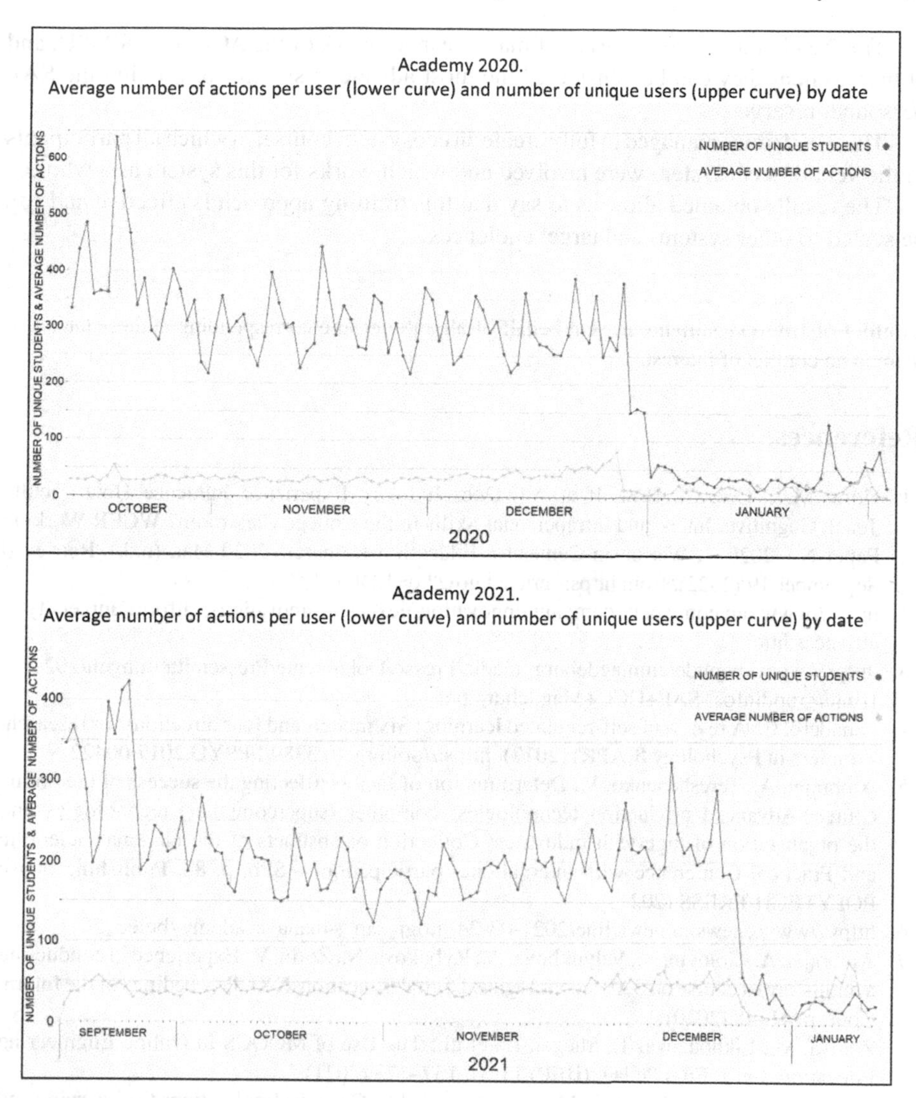

Fig. 10. Comparison of behavioral characteristics of the participants during the course

8 Conclusions and Key Outcomes

Based on the results of the two Academies, we can conclude that the participants showed very high performance results for the free open course.

We can talk about the stability and reproducibility of the pedagogical result [13].

Student surveys show a consistently high level of satisfaction with the courses themselves and with the additional content from partners.

The use of the online platform made it possible to effectively collect the individual digital footprint of students and analyze it to cluster students and identify factors that affect the results of their mastering of the course [5, 15].

The Academies were the result of many years of work of the ACC SAP SPbPU, and at the moment they can be considered the most advanced system for training the SAP personnel reserve.

The Academies managed to fully create an ecosystem course, in which all participants in the vendor's ecosystem were involved and which works for this system as a whole.

The results obtained allow us to say that this training approach is effective and can be scaled to other systems and target audiences.

Conflict of Interest Statement. On behalf of all authors, the corresponding author states that there is no conflict of interest.

References

1. Hora, M.T., Lee, C.: How, If at All, Does Industry Experience Influence How Faculty Teach Cognitive, Inter-, and Intrapersonal Skills in the College Classroom? WCER Working Paper No. 2020–2, Wisconsin Center for Education Research, 2020-Mar. (n.d.). Retrieved September 19 (2022). from https://eric.ed.gov/?id=ED603917
2. https://www.sap.com/about/company/innovation/next-gen-innovation-platform/university-alliances.html
3. .https://www.ovgu.de/unimagdeburg_media/Presse/Dokumente/Pressemitteilungen/2021/Hintergrundinfos+SAP+UCC+Magdeburg.pdf
4. Panadero, E.: A review of self-regulated learning: Six models and four directions for research. Frontiers in Psychology **8**(APR) (2017). https://doi.org/10.3389/FPSYG.2017.00422
5. Ambrajei, A., Tereshchenko, V.: Determination of factors affecting the success of the online course. Advanced production technologies: computer (supercomputer) technologies and the organization of high-tech industries: Collection of abstracts of the National Scientific and Practical Conference with international participation. – SPb, p. 88. Publishing house POLYTECH-PRESS (2021)
6. https://www.cnews.ru/news/line/2021-03-24_itogi_sap_s4hana_academy_bolee_50
7. Ambrajei, A., Golovin, N., Valyukhova, A., Rybakova, N., Zorin, V.: Experience in conducting a multi-format course on SAP technologies. Digital Education. XXI. Proceedings of the Intern. Conf. p. 41–48 (2020)
8. Zotova, M., Likhouzova, T., Shegai, L., et al.: The Use of MOOCS in Online Engineering Education. Int. J. Eng. Pedag. (iJEP) **11**(3), 157–173 (2021)
9. Engeness, I.: Tools and Signs in Massive Open Online Courses: Implications for Learning and Design. Human Development (0018716X). **65**(4), 221–233 (2021). https://doi.org/10.1159/000518429
10. Lyons, K.M., Lobczowski, N.G., Greene, J.A., Whitley, J., McLaughlin, J.E.: Using a design-based research approach to develop and study a web-based tool to support collaborative learning.Computers and Education **161** (2021). https://doi.org/10.1016/j.compedu.2020.104064
11. Liu, M., Zha, S., He, W.: Digital Transformation Challenges: a Case Study Regarding the MOOC Development and Operations at Higher Education Institutions in China. TechTrends: Linking Research & Practice to Improve Learning **63**(5), 621–630 (2019). https://doi.org/10.1007/s11528-019-00409-y
12. Ambrajei, A.N., Golovin, N.M., Valyukhova, A.V., Rybakova, Zorin, V.Y.: Use of hybrid learning model for SAP-related technology education. ICID-2019. http://ceur-ws.org/Vol-2570/paper10.pdf

13. Tarasova, O., Bykov, A., Likhouzova, T., Petunina, I.: Engineering Education and Learning outside the Classroom through MOOCs: Student Assessment. TEM Journal. **11**(1), 463–471 (2022). https://doi.org/10.18421/TEM111-59
14. Julia, K., Peter, V.R., Marco, K.: Educational scalability in MOOCs: Analysing instructional designs to find best practices.Computers and Education **161** (2021). https://doi.org/10.1016/J.COMPEDU.2020.104054
15. Ambrajei, A.N., Golovin, N.M., Valyukhova, A.V., Rybakova, N.A.: Using SAP Predictive Analytics to Analyze Individual Student Profiles in LMS Moodle. In: Taratukhin, V., Matveev, M., Becker, J., Kupriyanov, Y. (eds.) Information Systems and Design. ICID 2021. Communications in Computer and Information Science, vol 1539. Springer, Cham (2022). https://doi.org/10.1007/978-3-030-95494-9_6

Design Thinking Seminar for the Next Generation: Comparison of Short, Medium, and Long-Term Projects

Skrupskaya Yulia[1(✉)] and Victor Taratukhin[2]

[1] University of Lapland, Rovaniemi, Finland
yskrupsk@ulapland.fi
[2] SAP University Alliances, Palo Alto, USA

Abstract. Design thinking is a human-centered approach that revolutionizes the way you create new goods, services, procedures, and organizations. Design Thinking is widely used in education process to teach students about design thinking process and to create valuable projects that can boost innovations in the companies that support research at academic organizations. The question of our research is about the length of design thinking seminar. We study what is the most efficient length of the course to achieve the stated goals.

In this paper we observe the main information about design thinking seminars run at higher educational institutions. We study the examples of short-, medium- and long-term design thinking courses. The outcomes and comparison provided in the research show that design thinking teaching that can evolve into successful innovative startups is most efficient if it takes three or more months.

Keywords: Design Thinking · Academia Industry Collaboration · Innovation · Education

1 Introduction

Design thinking is a problem-solving approach that involves empathy, creativity, and experimentation to understand and address the needs of users [1]. It is often used in product and service design but can also be applied to other areas such as business strategy and organizational design. The goal of design thinking is to create innovative solutions that are user-centered and meet real needs [3].

Design thinking process usually involves five steps [1]:

1. Empathize: Understand the needs, wants, and limitations of the user through observation, research, and interviews.
2. Define: Clearly define the problem or opportunity that needs to be addressed.
3. Ideate: Generate a wide range of ideas and potential solutions through brainstorming and other idea-generation techniques.
4. Prototype: Create a physical or digital representation of the most promising ideas to test and refine them.

V. Taratukhin et al. (Eds.): ICID 2022, CCIS 1767, pp. 96–103, 2023.
https://doi.org/10.1007/978-3-031-32092-7_9

5. Test: Gather feedback on the prototypes and use it to iterate and improve the design.

It's worth noting that the process is often iterative, meaning that each step may be repeated multiple times and steps may be combined or re-arranged depending on the specific context. Additionally, some variants of design thinking may include additional steps or different terminology, but the five steps are the most common [4].

Design thinking is becoming increasingly popular in different fields and industries because of the number of benefits it has such as [2]:

- User-centricity. Design thinking puts the user at the center of the problem-solving process, ensuring that the solutions developed are tailored to their needs and preferences.
- Creativity and Innovation. Design thinking encourages creativity, experimentation and ideation, which allows to come up with innovative solutions to problems.
- Practicality. Design thinking is a hands-on approach to problem-solving that can be applied to a wide range of challenges, from product design to business strategy.
- Collaboration. Design thinking promotes collaboration and cross-disciplinary team-work, which helps to bring different perspectives and skills to the problem-solving process.
- Flexibility. The process of design thinking is adaptable and can be tailored to different contexts and situations.
- Iteration. Design thinking allows to test and refine solutions until they are optimized for the user.
- Empathy. Design thinking process starts with empathy, the ability to understand and appreciate the perspective of the user, which is crucial for creating user-centered solutions.
- Incremental change. It allows teams to test and validate solutions with minimal investment, before scaling up.
- Holistic approach. Design thinking is a holistic approach that considers the problem not just from one perspective but from multiple perspectives, which allows teams to identify the root cause of the problem and find a comprehensive solution [5].

All of these valuable advantages make design thinking a valuable skill for students to learn, and it's why it's becoming increasingly popular in universities as part of the curriculum [6].

Design thinking seminars are usually taught using the following strategies.

Start with a clear introduction to the design thinking process and its key principles, such as empathy and iteration. Then provide students with a design challenge or problem to work on, one that is relevant and relatable to their own experiences. It is also important to facilitate the brainstorming and ideation process by using techniques such as mind-mapping, sketching, and idea-generation exercises. Encourage students to create and test prototypes, this can be done with low-fidelity mock-ups, wireframes, or other types of physical or digital representations of their ideas and to gather feedback on their prototypes from their peers, instructors, and potential users [3]. Provide time for reflection and self-assessment, where students can reflect on their own design thinking process and strategies used. Encourage them to iterate and improve their designs based on the feedback they received. Incorporate real-world examples, cases, and guest speakers to provide students with a broader understanding of how design thinking is applied in different fields and industries. Encourage them to practice design thinking on different types of problems and industries. And of course, provide opportunities for students to share their work with others, through presentations, exhibitions, or other forms of public display [6].

The length of the design thinking process can vary depending on the specific project or problem being addressed. In general, it can take anywhere from a few days to several weeks or even months to go through the full process. Some specific steps within the process, such as prototyping and testing, may take longer than others. It's also worth noting that the process is often iterative, meaning that each step may be repeated multiple times and steps may be combined or re-arranged depending on the specific context. Also, the design thinking process is not a linear process, and it can be flexible, it can be adjusted to the complexity and scope of the problem and the resources available. In general, the key is to keep the process moving forward, and be flexible with the time frame, and make sure that you are not taking too long on one step while neglecting the others. The goal is to find a balance between spending enough time on each step to gather useful information and insights, but not so much time that the process becomes bogged down or loses momentum.

2 Design Thinking Seminars

In 2022 we participated in organizing several design thinking seminars that lasted for different periods of time. Seminars typically cover the key concepts and methods of design thinking, such as empathy, prototyping, and iteration. They may also include hands-on exercises and group activities to give participants the opportunity to practice the design thinking process. Design thinking seminars are a great way for individuals and organizations to learn about the design thinking process and how it can be applied to their own work. They can provide a valuable introduction to the key concepts and methods of design thinking and help participants develop the skills and mindset needed to use design thinking in their own projects. There is a brief description of each workshop below.

2.1 Short-term Design Thinking Seminar

Two-days elective for MBA and EMBA students of at School of Management Skolkovo. Usually this course is offline, but this year it took place in hybrid mode. The seminars were held in a high-tech auditorium with equipment that allowed you to conduct classes in the most modern formats. The equipment allowed the teacher to conduct a discussion with hundreds of listeners gathered in the audience, without being distracted by technical problems and maintaining eye contact with each participant. If the format of the event in the audience is mixed – online/offline – the professor connects remotely, and five cameras allow you to instantly "snatch" the speaker's face and transmit the picture to both him and other participants in Zoom. The professors and some other facilitators from the South Korea and the USA joined the seminar in Zoom, while students with assistants were present at the Business School auditorium. The majority of students were not familiar with the concepts of design thinking before the start of the seminar. Two-days seminar made it possible to familiarize students with the fundamentals of design thinking and practice its steps while creating projects in groups (Table 1, Fig. 1).

Table 1. Short-term Design Thinking Seminar

Duration	2 days
Dates	March 2022
Name of the course	Stanford Design-Based Method
Number of students	30
Audience	Top-level managers and specialists who are students of MBA and EMBA programs
Format	Blended learning
Initial goal	Teach students the basics of design thinking Run a workshop to show all steps of design thinking on practice Provide examples of real-world applications of design thinking to help students see the potential relevance of this approach for their current work
Results	The students unlocked their creative potential, leant how to create user experiences that resonate with the users and provide high strong business impact obtained

Fig. 1. Stanford Design-Based Method, 2022. (Source: personal archive)

2.2 Medium term Design Thinking Seminar

Design thinking seminar lasted for 3 weeks in total, with 4–6 h of face-to-face communication per week. It was a part of Internet Project Management course at the National Research University Higher School of Economics and showed students how they can use design thinking in project management. Majority of students were already familiar with the basics of design thinking (Table 2, Fig. 2).

Table 2. Medium term Design Thinking seminar

Duration	3 months
Dates	January – March 2022
Name of the course	Internet Project Management
Number of students	28
Audience	Master students all working mostly in IT companies
Format	Blended learning
Initial goal	Teach students how to use design thinking in project management students see the potential relevance of this approach for their current work
Results	The students leant how use Design Thinking methodology in project management, how to clarify project's goals and objectives, enhanced creativity, discovered how to pivot towards a project management style that ii user-centered and playful, includes empathy and creativity

Fig. 2. Design Thinking on Internet Project Management course, 2022. (Source: personal archive)

2.3 Long term Design Thinking Seminar

Stanford University ME310 university course that focuses on building solutions that anticipate the future. This programme was invented back in 1965. And it was the one that helped solve a bunch of different problems, creating many interesting approaches. In particular, ME310 has been used by Audi, IKEA, GM, Huawei, Lockheed Martin, Merck, Microsoft, Siemens and many others. The methodology involves complex tasks involving multicultural teams from different universities. Different ages, cultures, genders, specialties – all this gives the project a breadth of vision. This year students from the USA, Austria and Germany developed the solution of Yoi ecosystem that inspires employees and employers to enhance workplace wellbeing via wearable band to measure personalized wellness level. The decided to do this project because more than 70% of Americans report that they experience stress at work. Work-related stress has become one of the most serious health problems in the modern world. Long working hours and lack of resources are the main sources of stress. That is why it is important to lower the stress levels of people working in the companies.

About 80% of seminar participants of the design thinking seminars already knew the main concepts of design thinking while for another 20% it was a completely new approach. Students widely all the resources they were provided and at the end even received offer of 10000$ investment to continue their project (Table 3, Fig. 3).

Table 3. Long term Design Thinking seminar

Duration	10 months
Dates	September 2021 – June 2022
Name of the course	ME310 Future Talents project
Number of students	11
Audience	Master students with engineering major
Format	Blended learning
Initial goal	Teach students how to prototype and test their different design concepts Help them to create a full proof-of-concept system that demonstrates their ideas at the end
Results	The students leant how use Design Thinking methodology for creating startups, teamwork, international collaboration, received 10000$ funding for future development of the project from one of the partners

Fig. 3. ME310 Future Talents program participants working, 2022. (Source: personal archive)

2.4 Comparison

Stanford The best length of a design thinking workshop depends on the specific goals and objectives of the workshop and the level of familiarity of the participants with design thinking.

A short-term workshop can be a good introduction to the basics of design thinking, providing participants with an overview of the key concepts and methods. This type of workshop can be useful for participants who are new to design thinking and want to understand the basics of the process.

A medium-term course can be more in-depth, providing participants with more hands-on experience and the opportunity to practice the design thinking process in a more detailed way. This type of workshop can be useful for participants who are familiar with the basics of design thinking and want to develop their skills further.

A multi-day workshop provides more time for in-depth learning, practicing and refining the design thinking process. It also provides opportunities to work on real-world problems and to have more time for feedback and iteration. This type of workshop can be useful for participants who are already familiar with design thinking and want to apply it to a specific problem or challenge.

3 Conclusion

The best length of the design thinking process is the one that allows for a comprehensive understanding of the problem or challenge being addressed, the generation of a wide range of ideas, and the testing and validation of the most promising solutions. This length will vary depending on the specific project and the resources available as we saw in the examples discussed in this article. It's important to balance the need for a thorough understanding of the problem with the need for timely solutions. A short process didn't provide enough time to fully understand the problem and test them on greater number of users, in the medium term seminar there was enough time to do all steps of design thinking, but less team cohesion and no future project development, while a longer process led to delays but as the same more time provided teams travel and international collaboration opportunities which improved team cohesion, the quality of minimal viable product and the reality of continuing the project with investments.

To summarize, when facing the question about the length of the design thinking seminar it is important to look at the goals first. If the goal is to introduce students with design thinking method, then short- or medium-term seminar would be the best option. However, if the goal is to create a new innovative product or tool, medium- and long-term seminars would be the best solution.

References

1. Brown, T., Katz, B.: Change by Design: How Design Thinking Transforms Organizations, 272 p. Harper Publisher (2009)
2. Cote, C.: Why learn design thinking? Harvard Business School. Business Indights. [Electronic resource]. https://online.hbs.edu/blog/post/why-learn-design-thinking (2022)

3. Leifer, L., et al.: Engineering design thinking, teaching, and learning. J. Eng. Educ. **94**(1), 103–120 (2005)
4. Leifer, Lewrick, Link: The design thinking toolbox. [Electronic resource]. https://www.design-thinking-playbook.com/home-en?lang=en (2020)
5. Roger, M.: The Design of Business: Why Design Thinking is the Next Competitive Advantage, pp. 132–147. Harvard Business Review (2009)
6. Taratukhin, V., Pulyavina, N., Becker, J.: The Future of Design Thinking for Management Education. Project-based and Game-oriented methods are critical ingredients of success. Issue Vol. 47 (2020): Developments in business simulation and experiential learning. Volume: Innovations and Future Directions in Education, 18 Mar 03 2020

Accumulated Practical Experience of the Past for Building the Future of Design Innovations: Janus Project and Next-Gen Design Thinking Game

Natalia Pulyavina[2], Victor Taratukhin[1](✉), Seoyoung (Jenn) Kim[2], and Soh Kim[3]

[1] The University of Muenster, Leonardo Campus 3, Muenster, Germany
victor.taratukhin@ercis.uni-muenster.de
[2] Stanford Center at the Incheon Global Campus (SCIGC), Stanford University, Songdo, Incheon, South Korea
[3] Stanford University, Stanford, CA 94305, USA

Abstract. In this paper, we review the current state of the future of Design innovations including the Janus Initiative at Stanford and propose the active use of game-based methods as the important element of introducing the Next-Gen Design Thinking process, to understand the corporate history of innovation. Next-Gen Design Thinking (NG DT) is significantly extending the traditional Design thinking approach with a strong focus on the historical and cultural elements of participants. Next-Gen Design Thinking Innovation Game was developed based on Stanford Janus Initiative research as a tool for supporting the Next-Gen Design Thinking process for Engineering, Management, and IT students. Next-Gen Design Thinking Game supports a robust understanding Design thinking process, the corporate culture of industry partners, and innovation products history. Next-Gen Design Thinking Game implementation results will be compared with the traditional design thinking process. The way forward will be addressed, and the authors will propose future research directions.

Keywords: Design Thinking · Janus Project · Business Archaeology · Business artifacts · Innovation methods · Next-Gen Design Thinking · Next-Gen Design Thinking Game · Corporate Culture · Cultural elements

1 Introduction

1.1 Current Economic and Social Environment

The acceleration of social and economic changes and the impact of the latest events have yet to be fully understood. As always, in times of crisis and instability, there are growing demands for developing new approaches to the functioning of companies in challenging environments; thus, there is a need for a fresh look at the issues of designing and maintaining corporate innovations.

V. Taratukhin et al. (Eds.): ICID 2022, CCIS 1767, pp. 104–112, 2023.
https://doi.org/10.1007/978-3-031-32092-7_10

Design thinking, developed at Stanford University [1–3], is a powerful tool, and such a well-developed approach is actively implemented in global companies, leading business schools, and international organizations such as the UNDP [4].

It is also the number of research groups and consulting companies who are reviewing current Design thinking approaches and proposing corporate adoptions such as IBM [5], and SAP [6]. Despite heavy time investments to develop such adapted solutions, it is still a significant gap between the current DT process and customers' needs in a fast-changing, multicultural business environment [7].

1.2 Next-Gen Design Thinking

We believe it is impossible to form business creative practices of the future without experience of the Past, especially in the context of previous stages of the organization's Innovation process, a formed corporate culture, social and cultural components, and the spatial location of economic activities. Defining a new extended Design method will require taking a deep dive into the theory and practice of Archaeological science from one side and human-centered cultural experience from another side. As a result, Next-Gen Design Thinking (NG DT) [7, 8] is significantly extending the traditional DT approach with a strong focus on the historical and cultural elements of participants.

NG DT is a crucial approach to creating innovations specifically in corporate environments with strong corporate internal memory and cultural diversity. In addition, management, project-oriented education [7, 8], and game-based practices [7, 8] being embedded into the Next-Gen Design thinking process will enable engineering, IT, and management students to develop design projects and entrepreneurial skills while working in real-life innovation challenges formulated by an industry partner.

2 Business Archaeology and Janus Project

2.1 TAG Roundtable and Business Archaeology

Archaeology is a broad discipline that frequently reaches outside the boundaries of its academic setting. Its primary business application is attributed to its capacity for relating to objects, goods, materials, and design ideas of the past and transferring them to the present environment. However, an even greater question is what lessons can we take from our accumulated experience that may help us construct a better future?

The key ideas were discussed at TAG 2021 [9], Theoretical Archaeology Group Roundtable "The archaeological now: future world building" [10], by a diverse group of researchers from Stanford University, University of Muenster, Heidelberg University, academic institutions from the US, Denmark, UNESCO Future Literacy Group, Industry, and others. The roundtable discussed several key questions and provided valuable comments about developing archaeological methods in academic and business environments.

By using short, personalized case studies, the participants were able to hold a comprehensive discussion of this embeddedness amidst the contemporary experiences of exiles and emigres, the problems arising from the displacement of people and its impact on their identities, as well as growing tensions resulting from pervasive inequality and calls for social justice. The discussions included such actual topics as the perspectives of using archaeological methods and history in helping to understand and manage changes within communities, companies, and other enterprises; the possible correlation between archaeological hindsight and strategic foresight; the degree of influence of organizational history over its capacity for innovation and actual planning and performance; and, finally, the influence of individual or collective history and memories on shaping the common identity and the future prospects of a community or company.

2.2 Janus Project

The intense and productive discussion resulted in a new scientific initiative – Stanford Janus Project and Business Archaeology as a new Foundational Science [11].

The Janus Initiative [11, 12] seeks to facilitate human-centered design thinking and strategic foresight, aimed at stimulating business innovation and changes in corporate management. The capacity to innovate utilizes people's skills and competencies to study and understand basic and complex human needs and desires, which may result in completely new ideas and ways of presenting products, services, solutions, or experiences required in the future. The keystones of this approach are collaboration and communication – two concepts important for the survival and proliferation of mankind at all stages of history. And yet, design thinking still has not found an efficient way to incorporate memories and histories of the past preserved through various artifacts and elements of material culture into the way it can be used to create forecasts of the future.

The Janus Initiative employs a variety of well-established methods and tools that have proven their worth within design thinking and strategic foresight tool kit and skill set. Now, we only must use them to focus on the memories that are embedded in old products and buildings, in artificial landscapes and cityscapes.

On the other hand, archaeology has many ways of using the past embedded in sites and artifacts, for solving actual problems of the present day, i.e., providing a metaphorical bridge between the past and the future. As an example, we propose highlighting the past achievements of a company or community to show how it has managed to reach its current position, and how it has affected its psychological profile, namely, its capacity to change, adapt, and learn.

Museums and archives host myriads of artifacts and memories that can be a key to altering the self-image of a company or community to better fit the new vision of their future.

3 Next-Gen Design Thinking Game

3.1 Supporting the Design Thinking Process Through a Game-Oriented Approach

Next-Gen Design Thinking Game (NG DT Game) was developed for supporting the Next-Gen Design Thinking process for Engineering, IT, and Management students. NG DT Game was evaluated for international teams of students working for 3–9 months on design thinking challenges. Despite the limitations associated with COVID-19 and organizational difficulties, based on students' feedback NG DT Game has significantly helped in developing a deep understanding of the Design thinking process, the corporate culture of industry partners, and innovation products history.

Defining NG DT Game, we as members of Janus Initiative [11, 12] believe that the present day, the past, and the future meet. Like archaeologists, you will get treasures - artifacts from the past. The study of the artifact allows you to recreate the ideas and principles of doing business of the people who invented it, to feel the spirit of the era, the "archaeological layer". Then you will consider existing business development trends and create a new one in your field. This is exactly what archaeologists do: they study the past so that people can draw the necessary conclusions and create a better future, taking into account the achievements and mistakes of their ancestors. After that, you will strengthen and polish development with the help of powerful tools from today and with the active participation of other players.

Working in teams and team interaction helps to complement participants' development with unique experiences and vision of the problem by people from other areas of activity and get numerous insights. In addition to artifacts, you need Design thinking tools to support the extended Design thinking process and Trends definition to bring in an element of time and to make the process relevant to the current reality. We defined the Next–Gen Design Thinking Game concept below (Fig. 1) as the integration of Artifacts, Tools, and Trends cards.

Fig. 1. NG DT Game Concept. Artifacts, Tools, and Trends.

3.2 Next-Gen Design Thinking Game in Details

The NG DT Game set includes three types of cards: Artifacts, Tools, and Trends. The game can be used at various stages of the design thinking process. It is assumed that the teams are already working on some challenges. Each team draws three cards (one of each type). The teams then take turns presenting their ideas and prototypes and testing them with the help of other teams. Other teams give their feedback.

The cards inspire players to create new, broader ideas by immersing themselves in a different reality and getting acquainted with other experiences. The Artifact card talks about an extraordinarily successful or brilliant business idea from the past. The Trend card sets one of the existing trends within which the concept will be developed. The Tool card offers a specific method (from the Design Thinking and Archaeology arsenal) by which players optimize the concept they have developed.

When developing a concept, players are allowed to use the Internet to obtain additional information about the Artifact, Tools, or Trends. Artifacts, Tools, and Trends card examples are presented below (Fig. 2).

Fig. 2. An example of Trends (first row from the top), Tools (second row from the top), and Artifacts (third row from the top) in the NG DT Game.

Below is a prototype of a playboard that consists Design thinking process picture, specific zones for play cards associated with Design stages, and extra card zones (Fig. 3).

Some ideas about the way to implement a game-oriented approach to Design thinking have already been discussed [7, 8].

To summarize, the Next-Gen Design Thinking game has the potential in eliciting the cultural experiences of individuals, who are not only a designer but also an innovator or a creator willing to make a change. By applying these personal stories throughout a

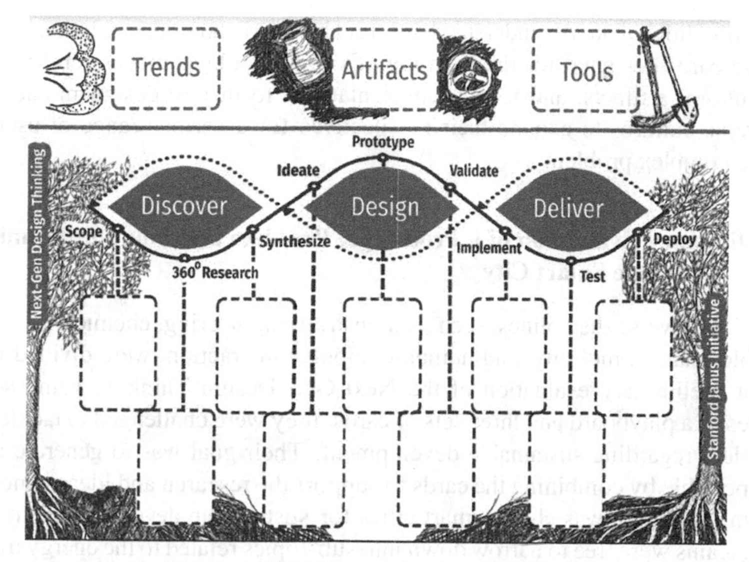

Fig. 3. NG DT Game play board template.

design thinking process, problems and solutions can be solidified and persuasive from the users' point-of-view. Incorporating cultural experiences adds value when it comes to generating a new idea or framing a new pattern. Applying one's cultural experiences might add new perspectives, which let designers immerse deeply into the users' scenario with empathy. Sharing a new cultural experience might inspire other team members who are unfamiliar with the culture, allowing them to indirectly imagine the situation or realize the differences. Combining diverse cultural aspects might trigger innovative ideas and stimulate creative thinking. Thus, depending on the users' circumstances, considering a specific cultural context or pre-analyzing human behaviors might become an essential step before framing the problem.

4 Incorporating Cultural Experiences

Design thinking tools, such as IDEO's Method Cards, SAP Scenes, SAP's Leporello, Board of Innovation's Brainstorm Cards, and many others from various organizations were created to keep a human-centered approach at the core of a design process [13–18]. According to several scholars, there are potential weaknesses of utilizing cards in a design process; but also, others illustrate the strengths and values of card decks as a supporter for "inspiration, organization, and communication of ideas" [18]. As there are many different design toolkits that focus on a single stage or an aspect of design process [17], each having a slightly different purpose, it is also important to develop a tool that can be applied in the entire design thinking process holistically.

The value of implementing cultural experiences into the design thinking process is proven to be significant as IDEO's method cards involve several culture-related contents [13]. For instance, the 'Cross-Cultural Comparisons' method from the learning category supports teams to compare different cultural groups when designing for an international market. This allows teams to understand different human behaviors in relation to the same topic. The cards designed for the Next-Gen Design Thinking process embrace global trends, cultural artifacts, and iconic representations to inspire designers and creators from diverse cultures to expand their perspectives for a various range of users while tackling a complex problem.

4.1 Preliminary Evaluation of a Prototype: Problem and Solution Framing for Sustainable Smart City

Experts from diverse disciplines, such as industrial engineering, chemical engineering, sustainable finance, mobility, and human-computer interaction, were divided into two teams for preliminary evaluation of the Next-Gen Design Thinking game set. With prototypes of a playboard and three sets of cards, they were challenged to tackle a smart city problem regarding sustainable development. Their goal was to generate as many ideas as possible by combining the cards to support the research and ideas generated by each team. The topic was about 'smart cities for sustainable development' in general; however, teams were free to narrow down into sub-topics related to the energy transition, carbon-free, mobility, or eco-friendly.

In the deck of cards, Korean traditional and contemporary artifacts are involved partially as this session took place in South Korea. The artifacts included Korea's region, monuments, trends, traditional objects and clothes, and food. These were selected based on the popularity of their use in Korean culture. We believe that personal and familiar experiences can strongly impact an individual's creativity in the sense of generating a unique idea to redesign a new concept for an existing problem (Fig. 4).

Fig. 4. Next-Gen Design Thinking game set (play board & cards)

Based on the processes and usage of cards, insights were found during the session. One team flipped the cards randomly from each category and tried to find a story connected to their own experiences related to the keywords. This sparked new ideas which later became the key to solving the issue in which they were tackling upon. Another team spread out all the cards at once to choose keywords that would inspire and support their ideas. As such, teams could engage more actively with randomness and unexpected connection of cards from different categories, which helped them to connect with their unique cultural stories and to use those personal experiences to think outside the box.

5 Implications and Conclusion

Active use of game-based methods within a design thinking process leverages the creativity of the teams allowing them to be more open-minded, collaborative, and immersive in the situation. Moreover, Next-Gen Design Thinking significantly extends the boundaries of the traditional design thinking approach with a strong focus on the historical and cultural elements of participants by recalling their unique experiences and memories that will impact their way of thinking throughout the problem-solving process.

This paper showcased a newly developed design thinking toolkit and explained how it can be used to implement designers' knowledge and cultural experiences for creating innovative outcomes in various contexts. In this regard, the limitation lies in receiving feedback from a small group of experts. However, in our prospective research, we plan to evaluate this toolkit by involving more participants from different countries and areas of expertise and collecting extensive feedback through conducting a survey and a short interview. Additional artifacts card sets development is underway for Central Asia (Uzbekistan) and the Nordic region (Finland). As such, more feedback regarding the guidance and improvement will allow us to develop the contents and structure of this toolkit that will match the needs of diverse players, who will use design thinking to overcome complex problems. We expect that Next-Gen Design Thinking will continuously develop to be more efficiently used in digital settings as well.

References

1. Leifer, L., Mabogunje, A., Sonalkar, N.: Design thinking: a new foundational science for engineering. Int. J. Eng. Educ. **32**(3), 1540–1556 (2016)
2. Liu, W., Byler, E., Leifer, L.: Engineering Design Entrepreneurship and Innovation: Transdisciplinary Teaching and Learning in a Global Context. In: Marcus, A., Rosenzweig, E. (eds.) HCII 2020. LNCS, vol. 12202, pp. 451–460. Springer, Cham (2020). https://doi.org/10.1007/978-3-030-49757-6_33
3. Ge, X, Leifer, L.: Design thinking at the core: learn new ways of thinking and doing by reframing. In: Proceedings of the ASME 2017 International Design Engineering Technical Conferences and Computers and Information in Engineering Conference, vol. 7: 29th International Conference on Design Theory and Methodology. Cleveland, Ohio (2017)
4. UNDP Design thinking. https://www.undp.org/arab-states/publications/design-thinking. Last accessed 3 June 2023
5. IBM Design thinking. https://www.ibm.com/design/thinking/. Last accessed 3 June 2023
6. SAP AppHaus,.https://apphaus.sap.com/approach. Last accessed 3 June 2023
7. Taratukhin, V., Pulyavina, N., Becker, J.: Next-Gen Design thinking. Using Project-based and Game-oriented approaches to support creativity and innovation. In: CEUR Workshop Proceedings, vol. 2570 (2020)
8. Taratukhin, V., Skrupskaya, Y., Kozlova, E.,Yudina, V., Pulyavina, N.: Bringing together engineering and management students for a project-based global ideathon: towards next-gen design thinking methodology. In: ASEE.2021, vol. Working Together: Approaches to Inclusivity and Interdisciplinarity (2021)
9. Theoretical Archaeology Group 2021. http://tag2021.stanford.edu/. Last accessed 3 June 2023
10. Theoretical Archaeology Group 2021: Workshop. The archaeological now: future world building. https://web.stanford.edu/group/tag2021/cgi-bin/wordpress/sunday-may-2/the-archaeological-now. Last accessed 3 June 2023

11. Shanks, M.: Janus Initiative. https://web.stanford.edu/group/archaeolog/cgi-bin/archaeolog/janus-initiative/. Last accessed 3 June 2023
12. JANUS Initiative Web site. https://janus.stanford.edu/. Last accessed 3 June 2023
13. IDEO Method Cards. https://www.ideo.com/post/method-cards. Last accessed 12 Jul 2022
14. SAP Design Thinking Leporello. https://apphaus.sap.com/resource/leporello. Last accessed 7 Dec 2022
15. SAP Scenes. https://apphaus.sap.com/resource/scenes. Last accessed 3 June 2023
16. Board of Innovation. https://www.boardofinnovation.com/tools/brainstorm-cards/. Last accessed 8 Dec 2022
17. Youngblood, M., Chesluk, B., Haidary, N.: Rethinking Users: The Design Guide to User Ecosystem Thinking. BIS Publishers, The Netherlands (2020)
18. Roy, R., Warren, J.: Card-based design tools: a review and analysis of 155 card decks for designers and designing. Des. Stud. **63**, 125–154 (2019)

Design Guidelines for Future Electric Vehicle Charging Stations

Heejung Yim[1(✉)], Seoyoung Kim[1], and Soh Kim[2]

[1] Stanford Center at the Incheon Global Campus (SCIGC), Stanford University, Incheon 21985, South Korea
{hyim,jennsy}@stanford.edu

[2] Civil and Environmental Engineering Department, Stanford University, Stanford, CA 94305, USA
sohkim@stanford.edu

Abstract. As the electric vehicle (EV) market is growing, the demand for convenient and fast charging is also increasing. As more people own electric vehicles (EV) and travel long distances with their EVs, more public charging infrastructure is required. This paper analyzes EV charging stations not only as sustainable charging hubs but also as recreational places for EV drivers and other users. In this research, we conduct a comprehensive literature review and case studies, focusing on how architectural design and associated services can play a pivotal role in encouraging more sustainable behaviors among users. To define the fundamental concept of the EV charging station, we explore the future concept and design of EV charging stations and qualitative data is gathered through interview-based cases to provide a solid understanding of user experiences. Finally, we propose design guidelines for creating sustainable and environmentally-friendly user experiences at public EV charging stations, aiming to benefit the local community from relaxation, recreation, and aesthetic perspectives.

Keywords: Sustainable Design · Electric Vehicle (EV) · Public EV Charging Station · Lifestyle · Mobility · Energy · Design Guidelines

1 Design for Public EV Charging Stations

1.1 Research Background and Purpose

Smart mobility has become more accessible and provides easy mobility services for citizens. Electric Vehicle (EV) and EV charging stations are characterized as future technologies of smart mobility. These mobility technologies are considered to meet the needs of people's changing lifestyles and create values for sustainable urban systems [1].

Need for Public Charging Station. As more people own electric vehicles, EV drivers require a convenient charging environment and public EV charging stations are major infrastructure for smart mobility services. In the European Union, the share of home charging is expected to reduce 40% by 2030 as more middle to lower income households purchase EVs [2]. Renters in the United States do not have home charging systems since rental-property owners have little incentive to invest in EV supply equipment [2].

V. Taratukhin et al. (Eds.): ICID 2022, CCIS 1767, pp. 113–121, 2023.
https://doi.org/10.1007/978-3-031-32092-7_11

Need for Fast Charging. Public fast charging stations are necessary to increase EV adoption in terms of decreasing the level of anxiety of EV drivers – encouraging drivers to convert from ICE (Internal Combustion Engine) to EV [3]. According to Villeneuve et al., solutions related to building fast charging stations and services were the most popular ones out of many other new ideas regarding EV charging while driving and parked [4]. Pilot project performed by utility TEPCO (Tokyo Electric Power Company) demonstrated that EV drivers' behaviors were changed significantly. In this test, EV drivers returned to the office with approximately 70% of battery remaining when the public charging station was not installed. However, after the public EV charging station was built, drivers returned to the office with approximately 30% charged [3] meaning that the level of anxiety have decreased due to easy accessibility. As EV owners are increasing, there is an urge to build public fast EV charging stations, especially for people who are expecting to travel long-distance.

Creating Values for the Environment. The experience of driving EVs and utilizing sustainably designed charging stations allows people to place a greater value on environmental sustainability in their life than ever before [5]. EV adoption is a symbol of expressing one's self-identity concerning environment and sustainability [6]. In this regard, public EV charging stations will challenge designers to build a sustainable mobility ecosystem and urban system for non-EV drivers as well as EV drivers. Environmental beliefs and user awareness of environmental issues affect the intentions to purchase EVs and encourage users to use renewable energy [7].

In this paper, we outline the key architectural design factors that need to be considered for EV charging stations and how they influence users' lifestyles and environmental values. By focusing on human needs within the space and sustainability for energy charging, decision makers will be able to adopt sustainable design and technologies for EV charging stations. Consequently, we aim to define the design factors and guidelines to build EV charging stations for creating a positive impact on improving people's sustainable lifestyle and actions toward a sustainable behavior.

1.2 Process and Methods

This paper presents a literature review and case studies that examine user preferences for charging stations and their services, as well as interaction between users and EV charging spaces. Data from individual's comments, articles on electric vehicles and charging experiences are collected and analyzed to understand the role of charging stations in the future. Also, interview data of current EV owners' needs are referenced to define design considerations for future EV charging stations. To investigate further, 20 existing and conceptual designs of EV charging stations are reviewed. Importantly, we have included the results of the EV charging station design competition in 2022, organized by Electric Autonomy Canada, to understand the architectural style, materials, and services of the future charging station [8].

In this study, we have adopted the framework from Design for Sustainable Behaviour (DfSB) [9] and the Fogg Behavior Model (FBM) [10]. The framework from DfSB referenced the FBM to explain how design strategy could be applied to sustainable mobility services. The FBM shows the relationship between motivation, ability, and triggers that represents how these factors influence design approaches and practices, which make people to support sustainable behaviors [10]. According to the FBM, when people gain enough of these three factors from the external environment, they can change their behaviors. Therefore, generating a new concept of sustainable EV charging station can play an important role for the three factors of the FBM [10]. However, only increasing the motivation cannot change EV drivers' behaviors properly. The drivers need motivation and ability to make sustainable decisions on their mobility services. Innovative design of EV charging stations encourage ability along with motivation by showing simple and clear messages on how to behave sustainably. Thus, we expect EV charging stations to increase sustainable behavior and decisions of users to benefit our society; moreover, build a sound mobility ecosystem.

2 Challenges for EV Drivers on Charging Electric Vehicles: Define Hurdles and Pain Points of Users

Effort to increase EV adoption is tackled by inconvenient perceptions of the current EV charging process and equipment. Long charging times, EV purchasing cost, and limited availability of charging stations are the challenges in urban environments [11]. Among the hindering factors toward the EV driving environment, we could find several major obstructions that drivers concern the most. We have found common inconveniences of charging EV by interviewing EV drivers and analyzing articles. In general, they commonly avoid long distance travel with EV and mostly charge from their familiar place. Detailed problems are as follows.

First, drivers experienced a lack of information on EV charging stations. They commented it is not easy to find adjacent charging stations when they are in unfamiliar locations and check availability of the charging equipment. In this regard, charging reservations using mobile apps is one of the functions drivers want to use. Also, users want to know what is nearby so they can plan what to do while waiting for the car to fully charge [12].

Second, drivers encounter usability issues of EV supply equipment (EVSE). There are complaints about EV charging cables and EVSE. Location of the charging cable is difficult to reach, cables are heavy to move, and the length of cable is short to reach the vehicle. Besides, over 5.2% of the 26,000 public EV chargers in the UK were broken and out of order [13].

Another usability issue is about the information displayed in EVSE and the payment system. Information displayed while charging is not user friendly. For instance, charging cost, charging amount (%), charging time is not provided all together and drivers are often confused to receive the information they need.

Lack of a unified payment system in public charging stations is another challenge to adopt EV. There are diverse payment service providers and EVSE adopts different types of payment methods. This increases complexity and inconvenience of the payment system. Despite all these barriers, we need to define target behaviors of users, consider how to motivate users, and trigger their actions to accelerate energy transition through electric vehicles.

3 Motivation for Sustainable Behavior: What Makes People Behave More Sustainably?

EV charging stations can play a role in motivating and triggering behavioral change. Behavior that we try to achieve is called target behavior according to the FBM. Defining the target behavior is the primary work to create motivation and triggers for behavioral changes of users. For charging stations, the target behavior is to make people more clearly aware of the problems of carbon emissions and purchase electric vehicles to accelerate sustainability of communities. When motivation is increased, it is more likely that people will perform the target behavior [10].

To motivate users to take actions for sustainable living, there are three elements of motivation from the FBM we need to consider for the charging stations: pleasure, hope, and social acceptance [10]. Aesthetic design, innovative technologies, and convenient services give pleasure to users, and it is an important attribute that attracts people to charging stations. Hope is a virtuous motivator and an effective means to persuade people to purchase electric vehicles and a smart mobility ecosystem. Using the charging stations, people are expecting that they are socially and environmentally recognized as a front runner of EV. There are several questions to specify the motivators and triggers for charging stations:

- What do EV drivers want to do during the charging time?
- What are the user requirements on the infrastructure for charging?
- Does this bring more people to EV charging stations?

In this regards, we clarify the major design concepts based on its functionalities. 1) **Re-charging**: recharging people's psychological needs with nature-friendly design and environments as well as their electric vehicles; 2) **Smart-charging**: providing easy, convenient and fast EV charging experiences using renewable energy; 3) **Fun-charging**: waiting time in the station turns out to be recreational and playful. Based on these functionalities, we created design guidelines for sustainable EV charging stations. Interview analysis and various online comments about EV charging experience regarding six components of services are recommended to enhance drivers' motivations of re-charging, smart-charging, and fun-charging (Table 1).

Table 1. Type of required services by users in EV charging stations

Driver Type	Vehicle Maintenance	Vehicle Cleaning	Food & Beverage	Shopping	Entertainment	Lounge
EV driver (Owner)	High	High	Medium	High	Low	High
EV driver (Rental)	Low	Low	High	High	High	High
PM driver	High	Low	High	High	Low	High
EV or PM driver for delivery	High	Low	Low	Low	Low	High

EV: Electric Vehicle
PM: Personal Mobility

4 Exploring Designs for Sustainability: Case Studies and Best Practices of EV Charging Stations

In this research, we explored design cases from different sources and identified key design elements for future EV charging stations. From the case studies, common elements for charging stations were found. Cases show that they minimize the environmental pollution of building the exterior and other infrastructure while maximizing comfort and convenience of users. As electric vehicles are known as the solution for reducing emissions for mobility [4], we believe that EV charging stations should also use renewable, reused, or recycled materials to construct a more sustainable space.

Sustainable architecture design started from an ecological perspective leveraging ecological functions conducted by nature [14]. Case studies we examined reflect this approach. From the cases, we found that designers are trying to create environmental ecosystems and enhance biodiversity in the EV charging space. The key principles of green architecture are: Sustainable Site Design; Water Conservation and Quality; Energy and Environment; Indoor Environmental Quality; and Conservation of Materials and Resources [14]. Thus, we have selected several elements from these principles to apply when considering the design of sustainable charging station, which focused on the local ecology and culture of the community as well as technology. Sustainable charging stations support environmental, social, and economic benefits. Energy transition to renewable energy is accelerated by the extended use of electric vehicles and charging stations.

Conceptual designs are referenced from the Design Competition Project of Electric Autonomy Canada in 2022 [8]. Key design factors identified from the cases are summarized in the following chapter (Fig. 1).

Fig. 1. Case study: Architectural styles, materials, and designs for EV charging stations [8]

5 Design Components for Sustainable EV Charging Station

This study identified 10 design components for EV charging stations. EV charging time and space delivered a negative experience for the users; however, by following the design components and check points suggested in this chapter, it could turn out to be a pleasant and refreshing time and place for the users to enjoy (Table 2).

Table 2. 10 Design components and check points to consider for building EV charging stations

No.	Design Component	Check Point
1	ACCESSIBLE	• Is the charging station noticeable while driving? • Is the charging station located in an easily accessible area? • Is charging supply equipment easily accessible in terms of location, availability, and charging cable? • Is it accessible by everyone (including people with disabilities)?
2	AESTHETIC	• Does the design and architecture provide motivation for sustainable lifestyle and time for refreshment? • Does the station have good view or a park nearby for users to take a walk or refresh? • Is the atmosphere and design of the space appealing enough to attract people for a revisit?
3	MODULAR	• Is the architecture designed in modular units and easy to build? • Are modular units adaptable to diverse environments and resizing for expansion?
4	NATURE-BLENDED	• Does the station blend well with the natural environment? • Is the station designed environmentally to connect people with nature?
5	RECREATIONAL	• Does the station provide opportunities for recreation and relaxation? • Are there shopping places, restaurants, cafes, etc. which offer diverse recreational services to satisfy users? • Is there a park or a garden nearby? • Can the station provide outdoor activities for families and travelers?
6	RENEWABLE	• Does the charging station utilize renewable energy sources for EV charging? • Are there any plans for an energy storage system (ESS) to be installed for surplus energy allocation? • Does the design of charging station seamlessly integrate necessary infrastructure for utilizing renewable energy?

(*continued*)

Table 2. (*continued*)

No.	Design Component	Check Point
7	SEAMLESS	• Do charging space and recreational space of the station connect seamlessly to represent unified and singular aesthetic, structural, and environmental identity? • Can the visitors move between charging and other service areas without encountering barriers?
8	SMART	• Does the design of charging station offer infrastructure embedded smart technology as key components, such as payment, charging and information display, etc.? • Does the charging station offer innovative and creative services using smart technology that users can experience while charging their EV or using other provided services?
9	SUSTAINABLE	• Are sustainable materials used to build the charging station and do they minimize environmental impact? • Do materials, architecture, and concepts of the station align with the sustainability goals of EV technology?
10	USABLE	• Are charging cables and plugs easily reachable and movable by drivers? • Does the station install enough fast-charging equipment? • Are variable charging speeds provided to drivers to meet their needs? • Is it easy to maintain charging equipment? • Do drivers feel comfortable reporting errors of the equipment? • Is it easy to pay without installing all different types of mobile apps?

6 Conclusion: Creating a Sustainable Energy Future

This research examines the components of sustainable and innovative design approaches for the EV charging station analyzing its functionalities, key features, and services provided from the station. To accelerate the energy transition toward a net-zero future, more EV charging stations need to be designed and operated sustainably. A sustainable EV charging station should deliver positive effects through social, cultural, and recreational activity. Additionally, its design should strengthen local features physically and aesthetically to improve users' freedom of movement [15]. As ecosystems of EV charging stations create new players, technologies, and business models [16] in sustainable methods, the goal of our design guidelines for sustainable EV charging stations is to support decision makers and stakeholders in the energy charging ecosystems.

In this study, we defined what users seek from EV charging stations and the values that charging stations provide to the users. As such, it is crucial to emphasize and deeply understand the needs of diverse users and stakeholders surrounding the EV industry. A prospective study on emphasizing users and generating new ideas for constructing a sustainable EV charging station will be continued based on the design guidelines suggested in this research. We expect that the findings in this paper could contribute to building a platform of new technology to achieve sustainable energy, which eventually benefits local communities and citizens in the future.

References

1. Dell'Era, C., Altuna, N., Verganti, R.: Designing radical innovations of meanings for society: envisioning new scenarios for smart mobility. Creat. Innov. Manag. **27**(4), 387–400 (2018). https://doi.org/10.1111/caim.12276
2. Gilleran, M., et al.: Impact of electric vehicle charging on the power demand of retail buildings. Adv. Appl. Energy **4**, 1–10 (2021). https://doi.org/10.1016/j.adapen.2021.100062
3. Maia, S.C., Teicher, H., Meyboom, A.: Infrastructure as social catalyst: electric vehicle station planning and deployment. Technol. Forecast. Soc. Chang. **100**, 53–65 (2015). https://doi.org/10.1016/j.techfore.2015.09.020
4. Villeneuve, D., Füllemann, Y., Drevon, G., Moreau, V., Vuille, F., Kaufmann, V.: Future urban charging solutions for electric vehicles. Eur. J. Transp. Infrastruct. Res. **20**(4), 78–102 (2020). https://doi.org/10.18757/ejtir.2020.20.4.5315
5. Axsen, J., TyreeHageman, J., Lentz, A.: Lifestyle practices and pro-environmental technology. Ecol. Econ. **82**, 64–74 (2012). https://doi.org/10.1016/j.ecolecon.2012.07.013
6. Rezvani, Z., Jansson, J., Bodin, J.: Advances in consumer electric vehicle adoption research: a review and research agenda. Transp. Res. Part D: Transp. Environ. **34**, 122–136 (2015). https://doi.org/10.1016/j.trd.2014.10.010
7. Lane, B., Potter, S.: The adoption of cleaner vehicles in the UK: exploring the consumer attitude–action gap. J. Clean. Prod. **15**(11–12), 1085–1092 (2007). https://doi.org/10.1016/j.jclepro.2006.05.026
8. Electric Autonomy Canada: The electric fueling station of the future: a design competition 2022. https://evcharging.electricautonomy.ca/awards2022/. Last accessed 1 Dec 2022
9. Lilley, D.: Design for sustainable behaviour: strategies and perceptions. Des. Stud. **30**(6), 704–720 (2009). https://doi.org/10.1016/j.destud.2009.05.001
10. Fogg, B.J.: A behavior model for persuasive design. In: Proceedings of the 4th International Conference on Persuasive Technology, pp. 1–7 (2009). https://doi.org/10.1145/1541948.1541999
11. Madina, C., Zamora, I., Zabala, E.: Methodology for assessing electric vehicle charging infrastructure business models. Energy Policy **89**, 284–293 (2016). https://doi.org/10.1016/j.enpol.2015.12.007
12. Forbes: The Lack Of EV Charging Stations Could Limit EV Growth, 5 May 2021. https://www.forbes.com/sites/prakashdolsak/2021/05/05/the-lack-of-ev-charging-stations-could-limit-ev-growth/?sh=562fdb246a13. Last accessed 10 Dec 2022
13. The Eco Experts: Why are so many electric vehicle chargers broken? 31 Oct 2022. https://www.theecoexperts.co.uk/blog/broken-ev-chargers. Last accessed 12 Dec 2022
14. Ragheb, A., El-Shimy, H., Ragheb, G.: Green architecture: a concept of sustainability. Proc. Soc. Behav. Sci. **216**, 778–787 (2016). https://doi.org/10.1016/j.sbspro.2015.12.075
15. Carra, M., Maternini, G., Barabino, B.: On sustainable positioning of electric vehicle charging stations in cities: an integrated approach for the selection of indicators. Sustain. Cities Soc. **85**(104067), 1–22 (2022). https://doi.org/10.1016/j.scs.2022.104067
16. Anthony Jnr, B.: Integrating electric vehicles to achieve sustainable energy as a service business model in smart cities. Front. Sustain. Cities **3**(685716), 1–12 (2021). https://doi.org/10.3389/frsc.2021.685716

Design Thinking in Designing Career Guidance Services for High School Students

Krasilnikova Svetlana(✉) and Yury Kupriyanov

Federal State Autonomous Educational Institution of Higher Education, National Research University "Higher School of Economics", Nizhny Novgorod, Russia
svkrasilnikova@edu.hse.ru

Abstract. This article examines the application of design thinking methodology in the design process of career guidance services for high school students, thereby revealing the specifics of this method. The author considers and comments on the main stages of design thinking and identifies design thinking tools for the development of career guidance services. The methods were selected for each stage of design thinking separately. Thus, a unique approach of service development was formed. Approach suggested by the authors consists of three stages with structured set of actions: the preparatory stage, the development stage, which includes the choice of software architecture and the choice of tools for development and the final stage of testing, bug fixes. Based on the elaborated approach a service operation model was also built with key success factors of the service implementation identified. Singavio platform and BPMN notation were used to create service operartion model.

Keywords: design thinking · career guidance service · Signavio

1 Introduction

Every year, a large number of boys and girls aged 17 to 18 years old begin to look for the use of their powers and abilities "in adulthood." At the same time, according to the results of the study "Problems of human resource development in the IT industry" from SAP, for the vast majority of high school students (87%), the choice of a future profession becomes random: it has little to do with professional orientation and is strongly associated with external factors [7]. Hence there are difficulties with further education and subsequent employment. The reasons are not only in the "closeness" of the labor market for the young and inexperienced, but also in the fact that the vast majority of high school students have very vague ideas about the modern labor market, existing professions, are unable to correlate the requirements of a particular field of professional activity with their individuality. Another reason is the influence of parents on the choice of a future profession. Given the difference in generations, very often the opinion of parents differs from the opinion of the child. Excessive parental control in relation to professional decision-making by adolescents leads to negative results. With the development of informatization, the strategy of preparing children for a conscious choice of

V. Taratukhin et al. (Eds.): ICID 2022, CCIS 1767, pp. 122–129, 2023.
https://doi.org/10.1007/978-3-031-32092-7_12

profession is changing, which leads to qualitative transformations in the content of the organization of career guidance work at school. One of the ways to modernize the system of vocational guidance of schoolchildren is the development of a modern career guidance service. The authors argue that service should be aimed not only at high school students, but also at parents. When designing a service, it is rational to use new approaches to collect requirements, identify insights (sometimes event not explicitly displayed by their subject) and facilitate decision-making. One of possible approaches with a good fit for the parameters stated above is the design thinking methodology. Design thinking is an iterative process that is more abstract in nature, its drivers are not logic, statistics and numerical models, but empathy, intuition, emotional insights, experimental models, and the desire for novelty [1].

2 The Technology of Creating a Career Guidance Service for High School Students Using the Methodology of Design Thinking as the Main Tool for Obtaining a Quality Product

The approach for service development is understood as the optimal sequence of steps that will lead to a high-quality end result [3]. Let's distinguish three stages of development, each of which contains a sequence of steps and a methodology to achieve the result.

The preparatory stage is an important link in the development process. Before proceeding to its content, let's turn to a brief excursion into design thinking. The concept of design thinking was summarized in 1980 by Professor of architecture Brian Lawson in the work "How designers think". Stanford University teachers have introduced the methodology into education [1]. A distinctive feature of this methodology is its simplicity and effectiveness in creating innovative solutions. Simplicity is described by the fact that the methodology can be applied in a variety of situations: starting from simple life and ending with a business strategy and establishing relationships in the company. CEO of IDEO and author of the book "Design Thinking in Business" Tim Brown talks about the methodology of design thinking as an approach to designing innovative solutions focused on people. It is based on the tools used by designers and is used to integrate people's needs, business needs and technological capabilities [2]. In design thinking, there is not always a specific "next step" to which you can proceed. Each phase of the methodology is unpredictable and can lead to many "next steps" as opportunities. This is the peculiarity. The Hasso Plattner Institute of Design at Stanford, commonly known as D.School, describes design thinking as a five-step process. The stages of design thinking do not have a strict sequence. Many designers can swap them, execute them in parallel, or repeat them when necessary. Each stage of design thinking is a separate tool that affects the entire project. Stanford School distinguishes five stages of design thinking: empathy, define, ideate, prototype, test. The first four components are just included in the preparatory stage of the development of the service, and the latter is allocated a separate stage. This approach is distinguished by its specificity, complete description of the stages and their number [5].

Before proceeding to the direct development of the service, you need to understand who this product will be interesting to, what their needs, desires, what worries them, to study the target audience with the help of empathy. To reflect all these points, it is logical to use the method of constructing an empathy map in the format of the consumer quadrant (Fig. 1). It reflects the experiences of potential users and their interests, and this greatly simplifies the further search for current problems.

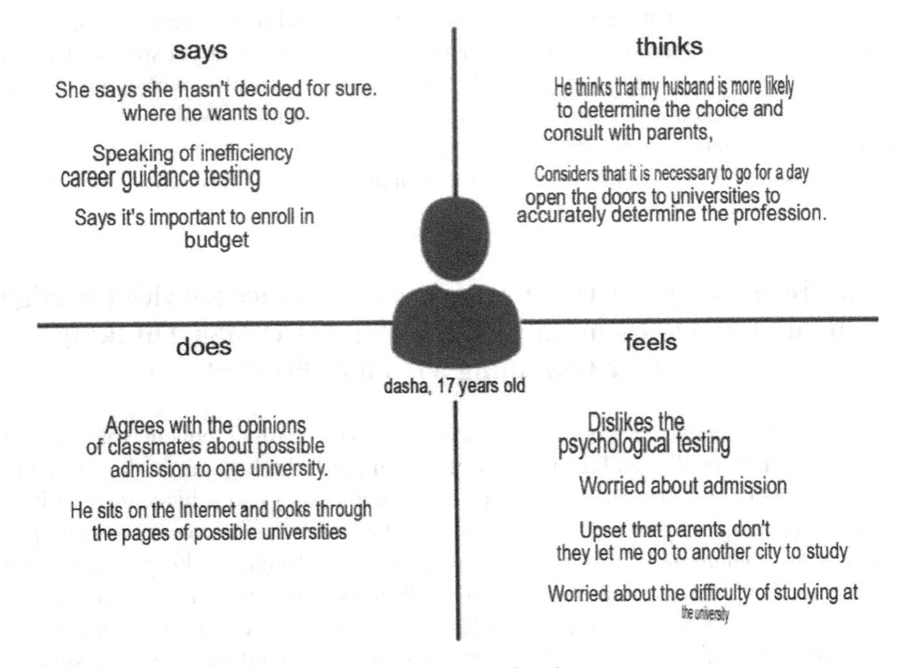

Fig. 1. Consumer Quadrant

After the facts about consumers are collected, we proceed to analyze and identify the problem that worries potential customers and generate ideas for its solution. To identify the problem, it is logical to use the "Why?" method. It allows you to build logical chains that lead to problems that users have. You can visualize this method using intelligence maps. (Fig. 2).

The result of the constructed intelligence map will be a formulated insight statement that describes the deep needs of a person and the problem he faces: I want to choose a profession that I would like, but I'm afraid that it will be difficult for me to study at university, that I will make a wrong choice and that we will have disagreements with parents. Insight explains all the desires and fears of the user. When the problem is defined, it is worth thinking about its solution. The "SCAMPER" technique will help in this. The "SCAMPER" technique is a scheme for posing certain questions that stimulate the generation of new ideas. In other words, it is a technique of creativity. Its author is Bob Eberle [4].

Before applying the SCAMPER method, it is necessary to clearly set the task: to identify a problem that needs to be solved, or an idea that needs to be developed. The use of the methodology in the context of the development of the structure of career guidance service is shown below (Table 1).

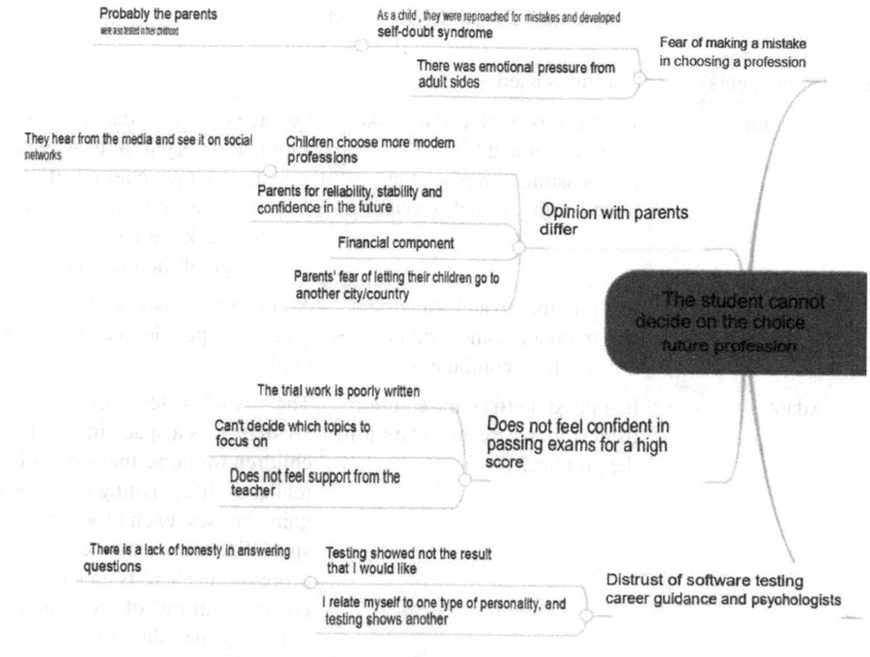

Fig. 2. Intelligence map

This method is universal and applicable to any problem. It focuses thinking on one part of the problem, which will allow you to quickly find the right idea. After all the ideas have been collected, it remains to create a single concept of the future service. Creating a prototype will help in this. The prototype is necessary to reflect the structure of the future service that the user sees. The prototype can be presented in the form of a flowchart, or created in a special application, for example, in Justinmind, Miro or Tilda. Before proceeding to the product development stage, you should create a technical task. The technical specification contains basic information about the product, approximate terms, scope of application, as well as the necessary requirements.

Before proceeding to the service development stage, it is necessary to think about the software architecture and the choice of development tools. In the process of studying the types of software architectures, it was decided to use a client-server three-level architecture. All client accesses to the database occur through the middleware, which is located on the application server. As a result, the flexibility of work and the productivity of the service increases. The choice of development tools is carried out jointly with the developer separately for the server and client parts. To develop the server part, you should decide on the programming language. According to the rating of programming languages, which was published on GitHub, the largest web service for hosting IT projects and their joint development, Java occupies a leading place in Back-End programming. Programming languages are rarely used in their pure form. Very often, development is carried out using frameworks. Frameworks are a pre–designed template

Table 1. The SCAMPER method

Letter	Components	Questions used	Ideas
S	Substitut	What can be replaced to make an improvement? Is it possible to replace the functionality, visual, service?	Replace career guidance testing with mandatory joint tests with parents, joint psychological consultations and a mandatory blog to check the level of knowledge of the material
C	Combine	Is it possible to add something else to create something new? Is it possible to combine objects?	Add podcasts, pages of successful people in a particular field
A	Adapt	Is it possible to take a solution from somewhere and adapt it to the product?	The city of professions "Kidburg" is a space in which children immerse themselves in real adult life, visiting numerous game houses, each of which specializes in a separate profession. There is an idea to create a "fitting" of professions – a 3-day marathon in which a high school student can experience professions for himself. The assistant will be a chatbot that acts as a boss. (Setting tasks for the day, issuing literature and assignments)
M	Modif	Is it possible to change part of the process or some functions to improve the product?	Remove the standard library of professions from the functionality, which may serve as a distracting aid
P	Put to Other Uses	Can it be used to solve other problems?	It can be used for employers to test new employees when hiring
E	Eliminate	Will removing any function help solve the problem?	Moving away from standard testing and the catalog of professions will allow you to look at the problem from a different angle
R	Rearrange	Is it possible to reverse an already existing concept?	Shift the focus from a career guidance service to a personal guide with built-in artificial intelligence for "fitting" professions

that allows you to speed up the development process by reusing its components. The most popular frameworks are presented below (Fig. 3).

Programming Language	Framework
Java	Spring, Apache Struts, Hibernate, Swing
PHP	Laravel, Yii, Zend
C#	.NET, UWP
Python	Django, Flask

Fig. 3. The most popular frameworks [8]

If the Java language is used for development, then the choice should be made in favor of Hibernate. This framework contains its own query creation language (HQL), which makes SQL queries simpler [3]. One of the criteria for choosing a database is the speed of access to data, specifically in a career guidance service to a selection of professions, skills, etc. Sometimes it is very important to view several professions at once in order to make the right choice, compare them with each other. And the speed of data access is determined by the amount of information viewed during the sampling process. Therefore, the higher the speed, the better the work on the site will be carried out. In relational databases, the process of accessing small data is fast. It is also easier to monitor the integrity of the data here. The most famous DBMS of this type will be Oracle, PostgreSQL, MySQL. The most popular Front-End development language is Java Script. As in Back-End development, frameworks are used in addition to a pure language. The most popular are React and Angular. Also, the main client part is written using HTML and CSS. HTML is responsible for the page frame, CSS is responsible for the page styling.

Product development is followed by a testing phase. The task of WEB testing is to find errors in the program and their possible elimination. The testing stage can be carried out in two levels: server and user. The main task of server testing is to check the equipment to meet all its characteristics: ease of use, simplicity, and efficiency. You can use special tools to test the server side. One of these is SQLBench, a tool that allows you to evaluate the work of a client-server application at an early stage of software development. The main task of user testing is to check the functionality of the service and ease of use for customers. At this stage, it is important not only to test the product, but also to track the emotions and feelings of the target audience.

For the visual operation of the service, it is recommended to build a model (Fig. 4). For modeling, a program for designing business processes – "Signavio" can be used and a BPMN diagram is selected. Business Process Model and Notation (business process modeling notation) is a system of symbols that displays business processes using flowcharts. The BPMN diagram shows the sequence in which work actions are performed and information flows are moved, and it is also the link between the analyst and the business manager.

After building the model, it will be correct to identify four main factors on which the success and effectiveness of the service implementation depends:

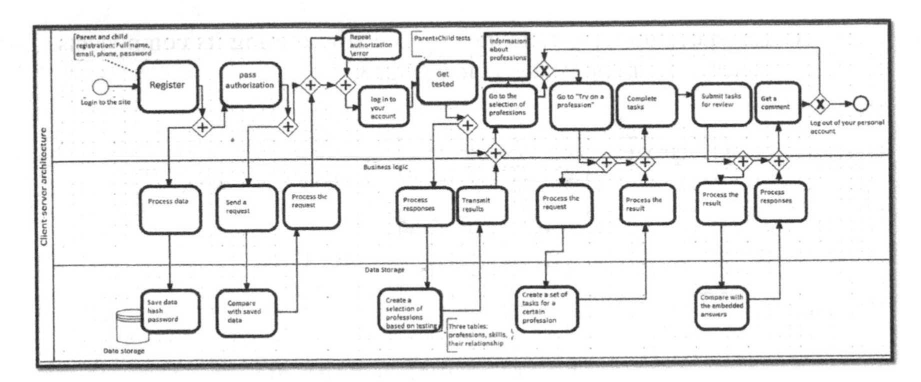

Fig. 4. Service operation model

- A systematic approach divided into separate stages: with this approach, developers can turn to the best experience and get the opportunity to avoid other people's mistakes and minimize their own;
- High-quality feedback: feedback allows you to identify shortcomings in the work of the service, as well as to fix new ideas that may come from respondents.
- User orientation: in order for an application or service to be in demand, it is necessary to find out the problem that the platform will solve, therefore, there is a need to study the target audience.
- Integration into existing infrastructures: this factor guarantees the rapid implementation of the project.

3 Results

As a result of going through all the stages of service development, a ready-made version of the product is obtained that can be used in reality. But we cannot exclude the fact that errors in the operation of the service may appear during its operation. Therefore, constant support of the site is necessary.

Thus, design thinking made it possible to create a high-quality career guidance service. Empathy allowed us to understand the experience and motivation of students, as well as immerse ourselves in the physical environment to gain a deeper personal understanding of the issues raised. The compiled empathy map has become a good tool for investigating the problem. In this case, it was the problem of the lack of a working career guidance service that benefits. At the stage of generating ideas, it was decided to create a service that will take into account all external factors that affect the choice of a child. In addition, the service gives you the opportunity to try yourself in any profession. This function allows you to quickly decide on the choice of a future profession. Next, a prototype of the service was created in order to understand what functions need to be added and what the interface will be. The next step in the development of my research will be testing the prototype. One of the easiest ways to test a prototype is to take a marketing survey. To make the survey more productive, all respondents will be divided into two groups. The first group will be shown the layout being developed, and the second group will be shown the layout of the site of the main competitor. Thus, shortcomings related to the structure of the site, the location of information and its design will be identified. According to the results of testing, the necessary adjustments will be made to the structure of the service.

References

1. Brown, T., Wyatt, J.: Design Thinking for Social Innovation, pp. 30–35. Stanford Social Innovation Review, Leland Stanford Jr. University (2010)
2. Brown, T.: Design Thinking in Business: From the Development of New Products to the Design of Business Models, 3rd edn, 256p. Tim Brown; translated from English, Vladimir Khozinsky. Mann, Ivanov and Ferber, Moscow (2018)
3. Kalyuzhny, E.R., Ksendzovskiy, I.D., Zarikovskaya, N.V.: Technology for Server Side Development Applications and Systems. Tomsk State University of Control Systems and Radioelectronics (2020)
4. Liedtka, J., Ogilvie, T.: Designing for Growth: A Design Thinking Tool Kit for Managers. Columbia Business School Publishing, New York, NY (2011)
5. Owen, C.L.: Design Thinking: Driving Innovation. A web article written for The Business Process Management Institute, 5 Sep 2006
6. Plattner, H., Meinel, C., Leifer, L.: Design Thinking – Understand, Improve, Apply. Springer (2010)
7. Zhilyaev, A.: Problems of human resources development in the IT industry of the Eurasian Union countries, p. 41c (2015)
8. Popular languages and frameworks for developing microservices: a large list: [сайт] List of popular languages and frameworks for developing microservices (mail.ru)

related to the [illegible] the [illegible] location of [illegible] and its [illegible] will be [illegible] results of testing. The necessary adjustments will be made [illegible]

[illegible]

[illegible]

[illegible] Systems and [illegible] 20[illegible]

[illegible] Thinking [illegible] Managers [illegible]

[illegible] of the Business [illegible]

[illegible] Apply. Springer [illegible]

[illegible] of [illegible]

[illegible]

Author Index

V. Taratukhin et al. (Eds.): ICID 2022, CCIS 1767, p. 131, 2023.
https://doi.org/10.1007/978-3-031-32092-7

Printed in the United States
by Baker & Taylor Publisher Services